Agroclimatology of Vegetable Crops Classification and Cultivation

NIPA® GENX ELECTRONIC RESOURCES & SOLUTIONS P. LTD.
New Delhi-110 034

About the Authors

BV Ramana Rao taught Agricultural Meteorology in the University of Agricultural Sciences, Bangalore during the years 1966 to 1979 and later joined the Indian Council of Agricultural Research and superannuated from service during April, 2000. He served as Head, Division of Climatology, Wind Power and Solar Energy Utilization at ICAR-Central Arid Zone Research Institute, Jodhpur (Rajasthan); founder Project Coordinator, All India Coordinated Research Project on Agrometeorology at ICAR Central Research Institute for Dryland Agriculture, Hyderabad (Telangana) during December 1984 to March 1994. His major contributions in the field of Agrometeorology include Agroclimatic Characterization, Delineation of Efficient Cropping Zones, Crop Weather Modeling, Classification and Management of Agricultural Droughts, Operational Agrometeorology, etc. He published about 150 research papers in peer reviewed journals with 50 of his papers appearing in high impact International Journals. He authored 15 books and monographs and a number of scientific and technical bulletins. He served as Editor in Chief, Journal of Agrometeorology during the years 2009 to 2023.

He is the recipient of British Council Fellowship during the year 1986-87 for advanced training at University of Reading, United Kingdom. He was conferred as Honorary Fellow (2011) and Life Time Achivement Award (2025) by Association of Agrometeorologists for his valuable contribution to the growth and development of Agricultural Meteorology in India.

Surender Singh has about thirty five years of experience with PhD in the field of Agrometeorology. Worked in Agrometeorology encompassing teaching, research, extension and administration at various capacities viz., Principal Scientist, Professor & Head, Advisor (Recruitments) and Associate Director (Students' Counseling and Placement) at CCS Haryana Agricultural University, Hisar, India. Recipient of Two Gold Medals for Best Research Work during PhD Program and for Best PhD Dissertation (2001-02) awarded by then President of India Dr APJ Abdul Kalam. Offered UK Universities Vice Chancellors' Scholarship for Higher Studies twice (1991 & 1993) under Commonwealth Fellowship Program. Post Doctoral Fellow at Abdus Salam International Centre for Theoretical Physics, Trieste, Italy;

IAC – Instituto Agronômico, Campinas, Brazil and BOKU (Met) - Institute of Meteorology and Climatology, Vienna, Austria.

He is life member of Association of Agrometeorologists, Indian Meteorological Society, Indian Society of Remote Sensing, International Society for Agrometeorology, Society for Sustainable Agric& Resource Management, Environment Science Published for Everybody Round the Earth (ESPERE), European Drought Centre, South Asian Forum on Agriculture Meteorology (SAFOAM) and The Geographical Society of India. Organized quite a few trainings/symposia in Agrometeorology and associated domains at National level. Published more than 250 research papers, popular articles, manuals/ teaching aids and research bulletins/book chapters etc. Performed responsibilities as Managing Editor - Journal of Agrometeorology, Research Collaborator, Monsoon Asia Agro-Environmental Research Consortium (MARCO), Japan; SPARC Research Collaborator, ETH, Zurich, Switzerland; Member, Expert Team - CAgM-WMO, Geneva, Switzerland (2018-20). Core Committee Member on Agriculture – Drought at WMO, Geneva, Switzerland (2021-24 & 2024-27). External Expert for evaluation of research projects submitted at COST-Association, European Union, Brussels, Belgium. Many students (UG, PG and PhD) were trained under him and placed across the country and well settled in various positions in the field of Agrometeorology. He is widely travelled researcher/expert across the globe on many academic pursuits. He has been conferred upon Fellow of Association of Agrometeorologists (FAAM) for the year 2024 in recognition of his exemplary achievements and enduring impact on agrometeorological science.

SK Dhankhar has thirty years of experience with PhD in the broader field of Vegetable Science. He is a distinguished Professor in the Department of Vegetable Science at Chaudhary Charan Singh Haryana Agricultural University, Hisar, India. With a robust academic background, he holds both MSc and PhD degrees in Agriculture, specializing in Vegetable Science. Dr Dhankhar has made significant contributions to the field of vegetable science, focusing on the impact of climatic factors on vegetable crop production. His research encompasses areas such as agronomy, horticulture, and the development of climate-resilient agricultural practices. He has been involved in studies assessing thermal requirements for phenological stages in crops like broccoli, aiming to optimize growth conditions and improve yield. In addition to his academic and research roles, Dr Dhankhar has served in administrative capacities. including the Project Director and Director Farm in the Directorate of Research in the university. Dr Dhankhar's work has been recognized in various scientific forums, and he continues to contribute to the advancement of vegetable science through research, education, and outreach activities. His efforts play a crucial role in developing sustainable and climate-smart agricultural practices in India and abroad.

V Uma Maheswara Rao has forty years of experience with PhD in the broader field of Agrometeorology. Worked in Agrometeorology encompassing teaching, research, extension and administration at various capacities viz., Project Coordinator, Principal Scientist, Professor & Head, Agrometeorologist, and Assistant Scientist at CCS Haryana Agricultural University and Indian Council of Agriculture Research in India.

He is life member of Indian Meteorological Society, The Arid Zone Research Association of India, Indian Society of Remote Sensing, Association of Agrometeorologists, International Society for Agrometeorology, and The Indian Society of Dryland Agriculture. Organized several trainings/workshops in Agrometeorology at the International and National level. Published more than 345 research papers, popular articles, manuals/teaching aids and research bulletins/book chapters. Member of the CCI-ET-IIC, WMO, Geneva (2014-2017). Member of XI Plan High Powered Review Committee of Ministry of Earth Sciences, Government of India (2012). In 2014, the prestigious Chaudhary Devi Lal Award, ICAR's Best AICRP on Agrometeorology was received under his dynamic leadership as Project Coordinator. Many students (UG, PG and PhD) have been trained under his guidance are placed across the country and settled in various positions in the field of Agrometeorology. He visited numerous countries in official capacity.

Agroclimatology of Vegetable Crops Classification and Cultivation

BV Ramana Rao
MSc Tech; FAAM
Founder Project Coordinator - AICRP on Agrometeorology
ICAR-CRIDA, Hyderabad, Telangana, India

Surender Singh
MSc; PhD; FAAM
Professor/Principal Scientist (Agrometeorology)
CCS HAU, Hisar, Haryana, India

SK Dhankhar
MSc; PhD
Professor/Principal Scientist (Vegetable Science)
CCS HAU, Hisar, Haryana, India

V Uma Maheswara Rao
MSc; PGDIAS; PhD; FAAM
Former Project Coordinator - AICRP on Agrometeorology
ICAR-CRIDA, Hyderabad, Telangana, India

NIPA® GENX ELECTRONIC RESOURCES & SOLUTIONS P. LTD.
New Delhi-110 034

NIPA® GENX ELECTRONIC
RESOURCES & SOLUTIONS P. LTD.
101,103, Vikas Surya Plaza, CU Block
L.S.C. Market, Pitam Pura, New Delhi-110 034
Ph : +91-11-43860225, Mob.: +91 9717133558, 9540816132
E-mail: newindiapublishingagency@gmail.com
Website: www.nipaersources.com

Print ISBN: 978-93-58874-19-8
ebook ISBN: 978-93-58876-40-6

Composed and Designed by NIPA®.

The Book

Agroclimatology of Vegetable Crops

Classification and Cultivation

is dedicated to

Professor Liberty Hyde Bailey

(1858-1954)

Professor LH Bailey was a pioneering American horticulturist, botanist, and educator who is widely recognized as the Father of Olericulture, the branch of horticulture that deals with the science, production, and marketing of vegetables. Born on a fruit farm near South Haven, Michigan USA, Liberty Hyde Bailey attended a small country school near the farm where he was born. He became a noted olericulturist, horticulturist, botanist, taxonomist, plant breeder, explorer, outstanding teacher, successful administrator, rural sociologist, prolific writer, editor, environmentalist, philosopher and visionary.

His pioneering contributions to vegetable science, tireless advocacy for rural life, and enduring influence on agricultural education have laid the foundation for generations of scientific inquiry and innovation in olericulture/horticulture. Professor Bailey is remembered not only as the father of olericulture but also as a visionary who transformed olericulture, horticulture, botany, and agricultural education. His influence persists through his writings, research institutions, and the generations of olericulturists he inspired.

As we explore the intricate relationships between climate and vegetable crops, we honor his legacy and draw inspiration from his lifelong commitment to the advancement of plant science and the well-being of those who cultivate the land.

Dr Tanja Likso

Head
Data Processing and Management Dept.
Croatian Met & Hydrological Service

Focal Point
National Climate Services Info Systems
World Meteorological Organization

Foreword

Vegetable crops form an indispensable component of global agriculture, playing a pivotal role in ensuring food security, nutritional well-being, and economic prosperity. Rich in vitamins, minerals, dietary fiber, and antioxidants, vegetables are crucial for a balanced diet and disease prevention, contributing significantly to human health. Their rapid growth cycle and high yield potential make them a vital source of income for millions of farmers worldwide, especially in developing nations.

The countries with diverse agro-climatic zones and a large agrarian population, vegetable cultivation holds immense significance. Beyond subsistence farming, vegetable cultivation has transformed into a highly commercialized enterprise, offering lucrative returns to farmers, generating employment across the value chain, and contributing substantially to the national agricultural GDP. The commercial importance extends from fresh produce markets to processing industries, encompassing frozen, canned, and dehydrated vegetables, as well as seed production.

However, the successful cultivation of vegetable crops is intrinsically linked to their interaction with the prevailing climatic conditions. This intricate relationship forms the core of agroclimatology, a discipline that studies the influence of climate on agricultural production. For vegetable crops, understanding agroclimatology is paramount, as each species and even variety has specific requirements for temperature, rainfall, humidity, light intensity, and photoperiod. Deviations from these optimal conditions can severely impact growth, yield, quality, and susceptibility to pests and diseases.

The increase in frequency of extreme weather events like droughts, floods, heat waves,changes in atmospheric CO_2 or ozone level considerably affect morphological, physiological and biochemical changes in the plants. The occurrences of these types of abiotic stress lead to membrane damage, oxidative burst, reduction of chlorophyll content and photosynthesis rate which have negative impacts on yields and quality of vegetable crops.

Despite the critical importance of climate in vegetable production, it is noteworthy that, until now, no serious or concerted efforts have been made to comprehensively understand and systematically document the intricate effects of weather parameters on vegetable crops. This gap highlights a pressing need for more intensive and dedicated research in this vital area. Such research is crucial for developing resilient cultivation strategies, adapting to climate change, and ensuring sustainable productivity in the face of increasingly variable weather patterns.

This book, "Agroclimatology of Vegetable Crops: Classification and Cultivation," aims to provide a comprehensive understanding of these critical aspects. It delves into the diverse classification of vegetable crops, not merely by botanical families but also by their climatic requirements, growth habits, and edible parts. A significant focus is placed on the seasonality of various vegetable crops, exploring how different species are adapted to specific seasons (rainy, winter, and summer) and how their cultivation practices must be adjusted to maximize productivity under varying environmental stresses.

By integrating principles of agroclimatology with practical cultivation techniques, this book serves as an invaluable resource for students, researchers, extension workers, and farmers. It is our sincere hope that the insights provided herein will empower readers to make informed decisions regarding crop selection, planting schedules, and management strategies, ultimately leading to enhanced productivity, improved crop quality, and sustainable vegetable farming systems. The authors Professors BV Ramana Rao, Surender Singh, SK Dhankhar and V Uma Maheswara Rao, who are well known for their contribution to education, research and development in their field of expertise globally, deserve appreciation for their valuable efforts in bringing out this important publication.

I am confident that the valuable book will definitely inspire researchers to explore the relationships between weather during different stages of growth and productivity of vegetable crops for developing climate resilient practices to achieve sustainable vegetable production.

(Tanja Likso)

Preface

India, one of the world's largest producers of vegetables, has achieved remarkable growth in vegetable production over the past few years. The latest statistics reveal a staggering production of over 193 million metric tons of vegetables in 2020-21, marking a significant increase from previous years. Despite this impressive growth, farmers cultivating vegetable crops continue to face formidable challenges posed by weather-related factors, including unseasonal rainfall, temperature fluctuations, drought, and extreme weather events. These climate-related stresses not only affect crop yields but also impact the quality and nutritional value of the produce, ultimately affecting the livelihoods of millions of farmers and the food security of the nation.

The impact of climate variability on vegetable production highlights the urgent need for a comprehensive understanding of the agroclimatology of vegetable crops. Unfortunately, the available information on this subject is scattered and fragmented, making it difficult for researchers, farmers, and policymakers to access and utilize it effectively. This book aims to bridge this knowledge gap by consolidating the available information on the classification and cultivation of vegetable crops, with a focus on their agroclimatic requirements.

The book is organized around the classification of vegetable crops based on seasonality, growth habits, and optimum climatic conditions required to achieve high productivity. It provides detailed information on the specific weather requirements of various vegetable crops, including temperature, rainfall, and solar radiation requirements. Additionally, the book examines the impact of extreme weather conditions, such as heat stress, frost, and drought, on the growth and productivity of vegetable crops. The book also highlights the best practices and strategies for mitigating the adverse effects of climate variability on vegetable production.

This book is a compilation of information from different sources, including research studies, government reports, and industry publications. We believe that this book will serve as a valuable resource for researchers, farmers, and policymakers working in the field of vegetable production, providing them with the knowledge and insights needed to improve the productivity and sustainability of vegetable crops in the face of climate change.

While this book provides a comprehensive overview of the agroclimatology of vegetable crops, we recognize that there is still a need for more intensive research on the subject, particularly in the context of climate change. We hope that this book will stimulate further research and innovation in this area and contribute to the development of climate-resilient vegetable production systems.

We are grateful to Dr Tanja Likso, Head-Data Processing and Management Department, Croatian Meteorological & Hydrological Service & Focal Point for National Climate Services Information Systems, World Meteorological Organization, Geneva, Switzerland for kindly providing a thought provoking and visionary foreword to this book, despite of his busy schedules. We hope that this book will make a meaningful contribution to the field of agroclimatology and vegetable production, and ultimately benefit the millions of farmers and consumers who depend on these crops for their livelihoods and food security.

BV Ramana Rao
Surender Singh
SK Dhankhar
V Uma Maheswara Rao

Acknowledgements

The successful completion of this book, **Agroclimatology of Vegetable Crops: Classification and Cultivation,** has been made possible through the support, collaboration, and encouragement of many individuals and institutions. We extend our deepest gratitude to the academic and research community for their valuable contributions to the field of agroclimatology and vegetable science, which have provided the foundation for this work.

We extend our heartfelt thanks to our colleagues at the Indian Council of Agricultural Research (ICAR), India Meteorological Department (IMD), World Meteorological Organization (WMO) and fellow researchers across agricultural universities and research institutions for their valuable inputs, critical reviews, and encouragement during the conceptualization and development of this book. Their multidisciplinary insights were essential in shaping the holistic approach reflected in the chapters.

We are particularly grateful to the farming communities, especially smallholder farmers, whose resilience and wisdom remain at the heart of this book. Their lived experiences and adaptive strategies have shaped much of the understanding presented in the book.

Authors are also grateful to the Association of Agrometeorologists (AAM) for their support and encouragement in advancing operational agricultural meteorology. The open source text material/images/references used in the book are also duefully acknowledged.

To our family and friends, their patience, understanding, and unwavering support have been the bedrock upon which this work stands.

We hope this book serves as a valuable resource for researchers, educators, students, extension professionals, and policy makers committed to climate-resilient vegetable farming.

BV Ramana Rao
Surender Singh
SK Dhankhar
V Uma Maheswara Rao

Contents

List of Figures

1

Unveiling the Climate-Vegetable Nexus for Sustainable Production

Vegetable crops stand as pillars of global food security and human health, providing essential vitamins, minerals, and dietary fiber crucial for well-being. Their diverse forms, flavors, and nutritional profiles contribute significantly to balanced diets worldwide. However, the successful cultivation of these vital crops is intricately linked to the surrounding atmospheric environment. This book, "Agroclimatology of Vegetable Crops: Classification and Cultivation" endeavors to explore and elucidate this critical relationship, providing a comprehensive understanding of how climate and weather influence the growth, development, yield, and quality of a wide array of vegetable crops. The interaction between vegetables and their climate is a dynamic and multifaceted one. From the moment a seed germinates to the final harvest, every stage of a vegetable plant's lifecycle is profoundly impacted by factors such as temperature, precipitation, solar radiation, humidity, and wind. Understanding these influences is no longer merely academic; in an era marked by increasing climate variability and the urgent need for sustainable agricultural practices, it has become a fundamental necessity for ensuring consistent and reliable vegetable production (Fig. 1.1).

Fig. 1.1: Vegetable crops harvest

This book aims to bridge the gap between the science of climatology and the art of vegetable cultivation. It delves into the specific agroclimatic requirements of major vegetable groups, including leafy greens, tuber crops, root vegetables, bulb crops, fruiting vegetables, cole crops, and legumes. For each category, we will explore the optimal climatic conditions, the detrimental effects of extreme weather events, and the ways in which growers can harness agroclimatic knowledge to make informed decisions.

Within these pages, readers will find detailed analyses of

- **Fundamental Climatic Parameters and Their Impact:** A thorough examination of how temperature regimes, rainfall patterns, solar radiation intensity and duration, humidity levels, and wind patterns affect key physiological processes in vegetable plants, such as photosynthesis, respiration, transpiration, and nutrient uptake.
- **Agroclimatic Zones and Vegetable Suitability:** Guidance on identifying suitable agroclimatic zones for specific vegetable crops, considering factors like growing degree days, frost-free periods, and water availability. This section will empower growers to select the right crops for their region, maximizing their chances of success.
- **Phenological Development and Weather Forecasting:** The book will explore the phenological stages of various vegetable crops and how weather patterns influence their progression. It will also highlight the utility of weather forecasting in planning critical agricultural operations, from planting & irrigation to pest & disease management and harvesting.
- **The Impact of Extreme Weather Events:** A dedicated focus on the adverse effects of climate extremes such as heat waves, cold spells, droughts, floods, and storms on vegetable yields and quality. We will also discuss mitigation strategies and protective measures that can be implemented to minimize these risks.
- **Microclimate Management in Vegetable Production:** Recognizing the significant role of the immediate environment surrounding vegetable plants, this book will detail various techniques for modifying microclimates within vegetable farms, including the use of mulches, row covers, shade nets, and windbreaks, to create more favorable growing conditions.
- **Climate Change Adaptation and Sustainable Practices:** Addressing the pressing challenges posed by climate change, we will explore adaptation strategies for vegetable production, such as selecting climate-resilient varieties, optimizing water use, and adopting sustainable farming practices that enhance soil health and reduce greenhouse gas emissions.

- **The Role of Technology in Agroclimatology of Vegetables:** An exploration of how modern technologies, including remote sensing, GIS, and data analytics, can be utilized to monitor weather patterns, assess crop health, and make precision agriculture decisions in vegetable cultivation.

1.1. Area and Production

Let's elaborate on the area and production of vegetable crops, looking at some more details and trends:

Global Perspective

- **China's Dominance:** As highlighted by multiple sources, China is by far the largest producer of vegetables globally. In 2022, their production was an astounding 616 million tons. This dominance is attributed to large cultivation areas, favorable policies, advanced agricultural technologies, and a diverse range of vegetables grown.
- **India's Significant Role:** India holds the second position in global vegetable production. In 2022, India produced 145 million tons of vegetables, showcasing its significant contribution to feeding a large population and also for exports.
- **Other Key Players:** The United States, Turkey, and Vietnam also feature among the top vegetable-producing countries, indicating the importance of vegetable farming across different continents and climatic zones.
- **Area Under Cultivation:** Globally, vegetables occupy a relatively small percentage (around 1.1% in 2010) of the total agricultural land. However, this land is often intensively cultivated, leading to substantial overall production.

1.2. India in Detail

- **Increasing Production:** India has shown a consistent increase in vegetable production over the years. Data from 2011 to 2022 indicates a positive trend in both the area under cultivation and the total production of vegetables.
- **Recent Estimates:** The second advance estimate for the year 2022-23 indicated a vegetable production of 212.91 million tonnes, showing a further increase from the previous year's 209.14 million tonnes. This growth can be attributed to various factors like government initiatives, improved cultivars and farming techniques, and expansion of cultivated areas.

- **Regional Contributions:** Different regions within India contribute variably to the total vegetable production. For instance, in 2016-17, the eastern region of India (including Bihar, Jharkhand, Odisha, and West Bengal) accounted for a significant share (around 33.72%) of the total area under vegetable cultivation and about 32.04% of the total vegetable production in the country. This highlights the importance of specific regions for vegetable supply.
- **Focus on Horticulture:** The horticulture sector, which includes both fruits and vegetables, is experiencing substantial growth in India, supported by initiatives like the National Horticulture Mission (NHM).

1.2.1. Factors Influencing Area and Production

- **Seasonality:** Vegetable production is highly influenced by seasons. Different vegetables thrive in specific temperature and moisture conditions, leading to variations in production throughout the year.
- **Market Demand:** Consumer preferences and market demand play a crucial role in determining which vegetables are grown and in what quantities.
- **Government Policies:** Government support through schemes, subsidies, and research initiatives significantly impacts the area and productivity of vegetable crops.
- **Technological Advancements:** Adoption of modern farming techniques, improved varieties, better irrigation methods, precision farming and protected cultivation (like greenhouses) contribute to higher yields and expanded cultivation in non-traditional areas.
- **Economic Viability:** The profitability of vegetable farming influences the area allocated to these crops by farmers.

While overall global vegetable production recently in 2022 was around 749 million metric tons (as per one source, with another stating 616 million tons specifically for the top 10 producing countries), here's a look at the production of specific vegetable types recently in the year 2022:

- ***Cucumbers and gherkins:*** *94.72 million metric tons*
- ***Yams:*** *88.26 million metric tons*
- ***Sweet potatoes:*** *86.41 million metric tons*
- ***Cabbages:*** *72.60 million metric tons*
- ***Eggplants (aubergines):*** *59.31 million metric tons*

These figures give insight into the production volume of some of the major individual vegetable crops globally.

1.2.2. Indian Vegetable Production

- **Top Producing States (Estimated 2023-24):** Based on data that appears to be preliminary or a specific category, the top producing states in India for 2023-24 (likely for a subset of vegetables) include:
  - Andhra Pradesh: ~321,600 units (Likely thousand tonnes)
  - Tamil Nadu: ~294,890 units
  - Maharashtra: ~279,278 units
  - Telangana: ~244,316 units
  - Karnataka: ~197,373 units
  - Madhya Pradesh: ~192,966 units
- **State-wise Production of Specific Vegetables (2021-22):** More detailed data for individual vegetables in 2021-22 (Production in thousand tonnes):
  - **Potato:** Uttar Pradesh (16808.74), West Bengal (11977.59), Bihar (7995.26)
  - **Onion:** Maharashtra (7316.37), Madhya Pradesh (6334.85), Gujarat (3719.45)
  - **Tomato:** Andhra Pradesh (1330.48), Madhya Pradesh (918.71), Karnataka (739.41)
- **Area under Specific Vegetables in India (2021-22):** (in thousand hectares)
  - **Potato:** Uttar Pradesh (471.79), West Bengal (413.41), Bihar (303.43)
  - **Onion:** Maharashtra (439.00), Madhya Pradesh (312.42), Gujarat (154.41)
  - **Tomato:** Andhra Pradesh (61.11), Madhya Pradesh (41.96), Karnataka (39.84)
- **Top Producing States (General):** Uttar Pradesh and West Bengal are consistently highlighted as leading states in overall vegetable production in India.
- **Production Trends in India:** India's vegetable production has shown a significant increase over the past decade, growing by approximately 32% between 2011-12 and 2023-24.

This detailed picture highlights the dominance of certain countries like China, India, USA etc. in global production and the significant contributions of states like Uttar Pradesh, West Bengal, Andhra Pradesh, Maharashtra, and Madhya Pradesh to India's vegetable output. The data also showcases the production volumes and cultivation areas for key vegetables like potato, onion, and tomato in India. Remember that these figures can vary slightly depending on the source and the year of data collection.

1.3. Economic Importance

The economic importance of vegetable production is very significant, contributing to livelihoods, national economies, and global food systems in numerous ways:

1.3.1. Contribution to Agricultural GDP and National Economy

- Vegetable production forms a substantial portion of the agricultural GDP in many countries. In India, with its large-scale vegetable cultivation, it plays a vital role in the overall economic output of the agriculture sector. Overall, this sector (horticulture and vegeatbles) contributes about 33% of the gross value to the agricultural GDP.
- The high frequency of cropping possible with many vegetables, along with their shorter growth cycles compared to fruits or field crops, allows for quicker returns on investment and contributes to a continuous flow of income in the agricultural economy.

1.3.2. Employment Generation

- The vegetable sector is a massive provider of employment, especially in rural areas. This includes:
  - **On-farm jobs:** Cultivation activities like planting, weeding, irrigation, fertilization, pest and disease management, and harvesting are labor-intensive and create numerous jobs.
  - **Post-harvest activities:** A significant workforce is engaged in sorting, grading, cleaning, packaging, transporting, and storing vegetables.
  - **Supply chain and marketing:** Wholesalers, retailers, and other intermediaries in the vegetable supply chain also provide substantial employment.

1.3.3. Source of Income and Livelihoods for Smallholders

- Vegetable farming is often a key source of income for small and marginal farmers due to the relatively smaller land requirements and potential for higher returns per unit area compared to traditional grain crops. This can significantly improve their economic well-being and contribute to poverty reduction in rural communities.

1.3.4. Nutritional Security and Public Health

- Economically viable vegetable production ensures the availability of affordable and nutritious food for the population. Vegetables are rich in vitamins, minerals, and fiber, contributing to better public health outcomes and reducing the burden on healthcare systems.

1.3.5. Support for Related Industries and Value Addition

- The vegetable sector fuels various allied industries:
  - **Food processing:** Vegetables are used in the production of frozen vegetables, canned vegetables, pickles, sauces, soups, and other processed food items, increasing their shelf life and market value.
  - **Seed and input industries:** The demand for high-quality seeds, fertilizers, pesticides, and irrigation equipment for vegetable farming supports these related industries.
  - **Packaging and logistics:** The transportation and packaging of fresh and processed vegetables create significant economic activity.

1.3.6. Export Earnings and International Trade

- Many countries with favorable climates and efficient production systems earn foreign exchange through the export of vegetables. Specific vegetables like onions, potatoes, and exotic vegetables are significant export commodities for some nations. Your earlier search indicated India's role in vegetable exports.

1.3.7. Regional Economic Development

- The concentration of specific vegetable crops in certain regions can lead to the development of specialized agricultural hubs, driving economic growth in those areas through infrastructure development, supporting businesses, and creating local markets.

1.3.8. Potential for Diversification and Higher Returns

- Vegetable farming offers opportunities for crop diversification, allowing farmers to grow a variety of vegetables throughout the year and cater to different market demands, potentially leading to higher overall returns compared to monoculture systems.

In conclusion, vegetable production is a critical component of the agricultural economy globally and in India. It provides widespread employment, supports numerous related industries, contributes significantly to national income, ensures nutritional security, and offers opportunities for economic growth and diversification, particularly for smallholder farmers. Understanding and supporting a robust vegetable production sector is therefore essential for sustainable economic development and food security.

1.4. Future Scenario

The future of vegetable production is poised for significant transformation, driven by several interconnected factors:

1.4.1. Increasing Demand and Production

- With a growing global population, the demand for vegetables will continue to rise. India, for example, anticipates a need for 225 million tons of vegetables by 2025 and 350 million tons by 2050, a substantial increase from the 184 million tons produced in 2017-18.
- To meet this demand with limited cultivable land, there's an urgent need to augment vegetable production through innovative techniques and technologies.

1.4.2. Impact of Climate Change

- Vegetable crops are particularly sensitive to climate variability. Changes in temperature and precipitation patterns can disrupt normal growth, flowering, pollination, and fruit development, ultimately decreasing yields (Fig. 1.2).

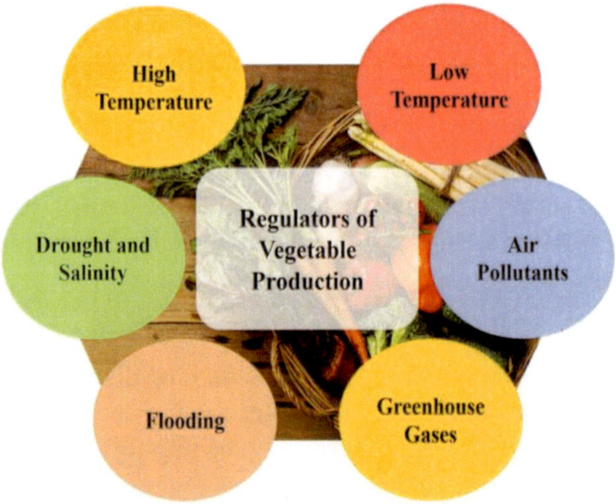

Fig. 1.2: Impact of climate change on vegetable production

- Extreme weather events like heat stress, droughts, floods, and frost will become more frequent, leading to crop failures and reduced quality, making vegetable cultivation more challenging and potentially unprofitable.

- Adapting to climate change through resilient varieties and mitigation strategies will be crucial for the future of vegetable production.

1.4.3. Technological Advancements

- **Precision Agriculture:** Expect widespread adoption of technologies like satellite imagery, AI-powered analytics, and sensors for precise irrigation, fertilization, and pest management, leading to optimized resource use and higher yields.
- **Automation and Robotics:** Robots capable of performing tasks like harvesting, weeding, and seeding with precision will help overcome labor shortages and improve efficiency.
- **Protected Cultivation:** Greenhouse and vertical farming will expand, especially in urban areas, enabling year-round production, higher yields per unit area, and reduced water and pesticide use. Vertical farming can increase yields significantly compared to traditional outdoor farming.
- **Biotechnology:** Development and use of genetically modified (GM) and gene-edited vegetables with enhanced traits like disease resistance, climate resilience, and improved nutrition will contribute to increased and more stable production.
- **Blockchain Technology:** This can enhance transparency and traceability in the vegetable supply chain, benefiting both farmers and consumers.

1.4.4. Sustainable Farming Practices

- **Regenerative Agriculture:** Practices focused on improving soil health, such as reduced tillage, cover cropping, crop rotation, and integrating livestock, will become more prevalent to enhance soil fertility and sequester carbon.
- **Integrated Pest Management (IPM) and Biocontrol:** A shift towards natural and biological methods for pest and disease control will reduce the reliance on chemical pesticides.
- **Water-Efficient Technologies:** Given increasing water scarcity, the adoption of efficient irrigation techniques like drip and micro-irrigation will be essential.

1.4.5. Changing Consumer Preferences

- Consumers are increasingly demanding high-quality, nutritious, safe, and sustainably produced vegetables.
- There is a growing interest in locally sourced and organic vegetables.
- Trends like plant-based diets will further drive the demand for a diverse range of vegetables.

The future of vegetable production will be characterized by a need to significantly increase output to feed a growing population while facing the challenges of climate change. This will necessitate the widespread adoption of technological innovations, a strong focus on sustainable farming practices, and responsiveness to evolving consumer demands. The vegetable sector is poised for significant growth and transformation in the coming years.

1.5. Role of Agroclimatology

The crucial role of agroclimatology in the future of vegetable production by diving deeper into each of the points mentioned earlier may be elaborated further:

1.5.1. Enhanced Climate Change Adaptation and Mitigation

- **Developing Resilient Varieties:** Agroclimatological data helps identify the specific climatic stresses that vegetable crops are likely to face in the future (e.g., increased frequency of heatwaves, prolonged droughts, altered rainfall patterns). This information is crucial for breeding programs focused on developing vegetable varieties with enhanced tolerance to these stresses. For instance, specific genes responsible for heat shock proteins or drought resistance can be targeted for introgression into popular varieties. Agroclimatological models can even predict which future climate scenarios these newly developed varieties will be best suited for.
- **Predicting Shifting Growing Regions:** As climate changes, the regions suitable for growing certain vegetables may shift. Agroclimatological models, incorporating future climate projections, can predict these shifts, allowing farmers and policymakers to plan accordingly. This might involve introducing new vegetable crops to regions that become climatically favorable for them or adapting existing farming systems in areas that become less suitable for traditionally grown vegetables (Fig. 1.3).

Fig. 1.3: Mitigation and adaptation strategies for vegetable crops to climate change

- **Guiding Mitigation Strategies:** Agroclimatology helps **to** understand the contribution of vegetable farming to greenhouse gas emissions, particularly through the use of fertilizers and irrigation. By analyzing the relationship between weather patterns and nutrient uptake, agroclimatological insights can guide farmers to optimize fertilizer application timing and rates, reducing the emission of nitrous oxide, a potent greenhouse gas. Similarly, efficient irrigation scheduling based on climatic data minimizes water usage and the energy required for pumping.

1.5.2. Precision Optimization of Resource Management

- **Water-Smart Irrigation:** Agroclimatological data, such as evapotranspiration rates (calculated based on temperature, humidity, wind speed, and solar radiation), provides precise information on the water needs of vegetable crops at different growth stages. This allows for the implementation of deficit irrigation strategies, where crops receive slightly less water than their potential requirement, optimizing water use without significantly impacting yield. Sensors in the field, guided by agroclimatic data, can further refine irrigation schedules based on real-time soil moisture levels and plant water status.
- **Nutrient Management:** Understanding the interplay between temperature, rainfall, and soil nutrients availability is critical for effective nutrient management. Heavy rainfall events, predicted by weather forecasts (an output of applied agroclimatology), can lead to nutrient leaching from the soil. Farmers can use this information to adjust fertilizer application timing or adopt slow-release fertilizers to minimize losses. Similarly, high temperatures can affect the rate of nutrient uptake by plants, requiring adjustments in fertilization strategies.

- **Optimizing Protected Cultivation:** In greenhouses and other protected environments, agroclimatological data from both outside and inside the structure is essential for maintaining optimal growing conditions. Sensors monitoring temperature, humidity, and light levels within the greenhouse can be controlled based on the specific requirements of the vegetables being grown and the external weather conditions, leading to energy savings and enhanced productivity.

1.5.3. Sophisticated Predictive Modeling and Early Warning Systems

- **Pest and Disease Forecasting:** Many pests and diseases of vegetable crops have lifecycles and infection rates that are strongly influenced by weather conditions. Agroclimatological models can integrate temperature, humidity, rainfall, **sun shine hours** and wind data to predict the likely emergence and spread of specific pests (e.g., aphid, whitefly, mites, *heliothes armigera*) and diseases (e.g., downy mildew, damping off, soft rot, leaf spot, little leaf, mosaic). This allows for timely and targeted interventions, reducing the need for broad-spectrum pesticide applications and minimizing environmental impact.
- **Extreme Weather Preparedness:** Agroclimatological forecasts, especially short-term and medium-range forecasts, are crucial for providing early warnings about impending extreme weather events like frost, heatwaves, or hailstorms. This enables vegetable farmers to take proactive measures such as covering vulnerable crops, providing shade, or adjusting irrigation to mitigate potential damage and losses.

1.5.4. Dynamic Site Suitability Assessment

- **Climate Change Impact Mapping:** Agroclimatologists use climate change projections and crop simulation models to create maps that show how the suitability of different regions for growing specific vegetables might change over time. This long-term perspective is vital for guiding agricultural land-use planning and investment decisions. For instance, a region that is currently ideal for growing tomatoes might become less suitable in the future due to increasing temperatures, prompting a shift towards more heat-tolerant crops or relocation of tomato production.
- **Considering Multiple Climatic Factors:** Site suitability assessments in the future will need to go beyond simply looking at average temperature and rainfall. Factors like the frequency of extreme heat events, the intensity of rainfall, changes in the length of the growing season, and even the impact of rising atmospheric CO_2 levels on vegetable physiology will need to be considered. Agroclimatological models are essential tools for integrating these complex interactions.

1.5.5. Tailored Microclimate Management

- **Optimizing Mulching Strategies:** Agroclimatological data can inform the choice of mulching materials and their application timing to achieve specific microclimate modifications. For example, in hot, dry regions, reflective mulches might be used to reduce soil temperature and conserve moisture, while in cooler regions, black plastic mulch might be preferred to warm the soil. The effectiveness of these techniques is directly linked to the prevailing weather conditions.
- **Designing Effective Row Covers and Shade Nets:** The type and timing of using row covers and shade nets can be optimized based on agroclimatological forecasts. Row covers might be used to protect against frost events predicted by short-term forecasts, while shade nets can be deployed during periods of intense solar radiation and high temperatures, as indicated by seasonal climate data.

1.5.6. Integrated Decision Support Systems

- **AI-Powered Farm Management:** The future will see the development of sophisticated decision support systems that integrate real-time weather data from local weather stations or satellite imagery with crop growth models, pest and disease prediction models, and even market information. These systems, powered by AI and machine learning, will provide vegetable farmers with highly personalized and timely recommendations delivered through mobile apps or online platforms. This could include advice on the optimal planting date based on historical weather patterns, precise irrigation schedules based on current and forecasted conditions, and alerts about potential pest outbreaks based on agroclimatic triggers.

1.5.7. Fostering Sustainable Production Ecosystems

- **Climate-Smart Crop Rotations:** Agroclimatological understanding can guide the design of crop rotation systems that are well-suited to the local climate and help improve soil health, reduce pest and disease pressure, and enhance water use efficiency. For instance, incorporating cover crops that thrive in the specific climate can improve soil structure and water infiltration.
- **Agroforestry Integration:** In some regions, integrating vegetable farming with trees (agroforestry) can offer microclimatic benefits such as shade and wind protection. Agroclimatological analysis can help determine the optimal tree species and spatial arrangement to maximize these benefits for vegetable production in a given climate.

In essence, agroclimatology will be the indispensable science that empowers vegetable producers to navigate an increasingly complex and challenging environment. By providing a deep understanding of the intricate relationships between climate and vegetable crops, it will pave the way for more resilient, efficient, and sustainable food production systems in the future.

1.6. Importance of Agroclimatology in Vegetable Production

Following our detailed discussion about the profound influence of agroclimatology on vegetable production (Fig. 1.4), especially considering the future challenges and opportunities, the need for a dedicated book on the "Agroclimatology of Vegetable Crops: Classification and Cultivaton" is clearly evident here:

Fig. 1.4: Agroclimatology for vegetable crops production

1.6.1. Centralized and Comprehensive Resource

- Currently, information regarding the intricate relationship between climate and vegetable crops is scattered across various research papers, agricultural guides, meteorological reports, and online resources. A dedicated book would consolidate this knowledge into a single, authoritative, and easily accessible resource for students, researchers, extension specialists, policymakers, and vegetable growers.

1.6.2. In-depth Crop-Specific Analysis

- Vegetable crops are incredibly diverse, each with unique climatic requirements and responses. A specialized book could provide a detailed analysis of the agroclimatology of major vegetable categories (leafy greens, root vegetables, bulb crops, fruiting vegetables, etc.) and even specific crops within those categories. This in-depth focus would allow for nuanced understanding and tailored recommendations that general horticulture books might not offer.

1.6.3. Practical Applications for Sustainable Production

- The book would serve as a practical guide for applying agroclimatological principles in real-world vegetable farming. It could provide step-by-step methodologies for:
  - Selecting appropriate vegetable crops and varieties for specific agroclimatic zones.
  - Utilizing weather forecasts and seasonal climate predictions for planning planting schedules, irrigation, and harvesting.
  - Implementing efficient irrigation and nutrients management strategies based on climatic conditions.
  - Developing integrated pest and disease management plans informed by weather patterns.
  - Employing microclimate modification techniques to optimize growing environments and protect crops from extreme weather.

1.6.4. Addressing the Urgency of Climate Change Adaptation

- With climate change posing a significant threat to food security, a dedicated book could focus on adaptation strategies for vegetable production. This would include:
  - Identifying and promoting climate-resilient vegetable varieties.
  - Exploring innovative farming systems (e.g., protected cultivation, agroforestry) that can buffer against climate variability.
  - Providing guidance on water-wise farming techniques for drought-prone regions.
  - Utilizing climate projections for long-term planning and risk management.

1.6.5. Integrating Modern Technologies

- The book could effectively integrate the role of modern technologies in the agroclimatology of vegetable crops. This would encompass:
  - Utilizing remote sensing and GIS for large-scale climate and crop monitoring.
  - Applying weather modeling and forecasting tools for improved decision-making.
  - Leveraging data analytics and artificial intelligence to identify climate-vegetable relationships and optimize farming practices.
  - Exploring the use of mobile apps and online platforms for delivering agroclimatic information to farmers.

1.6.6. Educational and Research Advancement

- A comprehensive book on this topic would be an invaluable educational resource for agricultural students and researchers. It could provide a strong theoretical foundation while also highlighting current research trends and future directions in the field. It could also inspire further research and innovation in developing climate-smart vegetable farming systems.

1.6.7. Bridging the Gap Between Science and Practice

- By presenting complex scientific concepts in an accessible and practical manner, the book could effectively bridge the gap between agroclimatological research and on-farm practices. This would empower vegetable growers to make more informed decisions, leading to increased productivity, improved resource use efficiency, and enhanced sustainability of their operations.

In essence, a dedicated book on the agroclimatology of vegetable crops is not just desirable but a necessity in today's agricultural landscape. It would serve as a cornerstone resource for building a more resilient and sustainable future for vegetable production, crucial for ensuring food security and healthy diets for a growing global population in the face of a changing climate.

1.7. Our Endeavor

This book provides a comprehensive and in-depth exploration of the critical relationship between climate and vegetable production. Its primary objectives are to:

- Establish a strong foundation in the principles of agroclimatology as they apply to the cultivation of diverse vegetable crops.
- Detail the specific climatic requirements and responses of major vegetable groups and selected individual species.
- Equip readers with practical knowledge and methodologies for utilizing agroclimatic information to optimize vegetable farming practices.
- Address the significant challenges posed by climate change to vegetable production and offer adaptation and mitigation strategies.
- Highlight the role of modern technologies in advancing the understanding and management of climate-vegetable interactions.
- Serve as a valuable resource for a broad audience involved in vegetable agriculture, from students and researchers to farmers and policymakers.

1.7.1. Target Audience/Stakeholders

This book is designed to cater to a wide range of individuals interested in the science and practice of vegetable production in the context of climate:

- Undergraduate and postgraduate students in agricultural sciences, olericulture/horticulture, environmental science, and related disciplines.
- Researchers and scientists working in the fields of agroclimatology, olericulture/horticulture, crop physiology, and plant pathology.
- Agricultural extension specialists and advisors providing guidance to vegetable farmers.
- Vegetable growers (both conventional and organic) seeking to enhance their understanding of climate impacts and improve their farming practices.
- Policymakers and government officials involved in agricultural planning, climate change adaptation, and food security initiatives.
- Professionals in the agricultural industry, including seed producers, input suppliers, and consultants.

1.7.2. Key Features and Content

The book is structured to provide a logical progression of knowledge, covering the following key areas:

- **Fundamentals of Agroclimatology:** An introduction to basic meteorological parameters and their relevance to plant growth and development, including temperature, radiation, precipitation, humidity, wind, and atmospheric composition.
- **Agroclimatic Zones and Vegetable Distribution:** Analysis of different agroclimatic zones around the world and their suitability for various vegetable crops, considering factors like growing degree days, frost dates, and water availability.
- **Impact of Climate Change on Vegetable Production:** A comprehensive assessment of the current and projected impacts of climate change, including temperature increases, altered precipitation patterns, extreme weather events, and rising CO_2 levels, on vegetable yields, quality, and pest/disease incidence.
- **Microclimate Management Techniques:** Practical strategies for modifying the microclimate in vegetable fields and protected environments (e.g. mulching, row covers, shade nets, windbreaks, greenhouses) to optimize growing conditions.

- **Water Management in Vegetable Production under Different Climates:** Principles and practices of efficient irrigation scheduling based on agroclimatic data, including evapotranspiration models and water-saving technologies.
- **Nutrient Management in Relation to Climate:** Understanding how temperature and rainfall affect nutrient availability and uptake in vegetable plants, and strategies for optimizing fertilization.
- **Climate-Smart Pest and Disease Management:** Utilizing agroclimatic information for predicting pest and disease outbreaks and implementing integrated management strategies to minimize losses.
- **Climate Change Adaptation Strategies for Vegetable Farming:** Exploration of resilient vegetable varieties, diversification of cropping systems, conservation agriculture practices, and other approaches to enhance the adaptive capacity.

1.7.3. Overall Aim and Expected Impact

This book aims to empower readers with a deep understanding of the intricate relationship between climate and vegetable crops. By providing a comprehensive resource on the principles, practices, and technologies in this field, it intends to contribute to:

- Enhancing the productivity and sustainability of vegetable farming systems.
- Improving the resilience of vegetable production to climate variability and change.
- Ensuring a more stable and secure supply of nutritious vegetables for a growing global population.
- Promoting environmentally responsible and economically viable vegetable farming practices.

The book is intended for a diverse audience, including agricultural students, researchers, extension specialists, vegetable growers (both conventional and organic), and policymakers involved in agricultural development and climate resilience. Our aim is to provide a comprehensive yet accessible resource that empowers readers with the knowledge necessary to understand and effectively manage the complex interplay between climate and vegetable production.

By embracing the principles of agroclimatology, we can move towards more efficient, sustainable, and resilient vegetable farming systems that contribute to enhanced food security, improved farmer livelihoods, and healthier diets for all. This book serves as a guide on that essential journey, illuminating the path towards a climate-conscious future for vegetable agriculture.

2

Agroclimatic Requirements for Fruit Type Vegetable Crops Grown in India

India's diverse climate and geography provide a favorable environment for growing a wide range of fruit-type vegetables, including eggplant, tomato, okra, and chillies. These crops are not only essential for food security but also contribute significantly to the country's agricultural economy. To optimize the production and productivity of these crops, it is crucial to understand their agroclimatic requirements.

Fruit-type vegetables have specific temperature requirements for growth and development. For instance, eggplant and okra require warm temperatures between 20-30°C (68-86°F) for optimal growth, while tomatoes prefer a slightly cooler temperature range of 15-25°C (59-77°F). Chilli peppers, on the other hand, thrive in temperatures between 20-35°C (68-95°F). Adequate moisture is also essential for the growth and development of fruit-type vegetables. Eggplant, okra, and chilli peppers require consistent moisture, especially during the fruiting stage, while tomatoes require a slightly drier soil condition. The ideal annual rainfall for these crops ranges from 600-1,200 mm, with supplemental irrigation provided during periods of drought.

Weather conditions, such as sunlight, wind, and frost, also play a crucial role in the growth and development of fruit-type vegetables. Most fruit-type vegetables require full sun to produce well, while excessive wind can damage plants and reduce yields. Frost, on the other hand, can be detrimental to these crops, and frost-free periods are essential for optimal growth. India's diverse climate and geography result in regional variations in agroclimatic conditions. For instance, the Indo-Gangetic Plains, which cover parts of Uttar Pradesh, Bihar, and West Bengal, have a subtropical climate with hot summers and cold winters. In contrast, the Deccan Plateau, which covers parts of Maharashtra, Karnataka, and Telangana, has a tropical climate with warm summers and mild winters. Understanding these regional variations is essential for optimizing the production and productivity of fruit-type vegetables. The agroclimatic conditions suitable for growing different fruit type vegetables are given in this chapter.

2.1. Eggplant (*Solanum melongena*)

Eggplant, also known as brinjal, is a significant vegetable crop in India, providing livelihoods for millions of farmers and contributing substantially to the country's food security (Fig. 2.1). Here are the details on the area and production of eggplants in India and their importance:

2.1.1. Area and Production

India is the second-largest producer of eggplants globally, accounting for over 25% of global production. According to the Ministry of Agriculture and Farmers' Welfare, the total area under eggplant cultivation in India is approximately 5.5-6.0 million hectares. The country produces around 13-14 million metric tons of eggplant annually, with major producing states including Maharashtra, Gujarat, Uttar Pradesh, and West Bengal.

Fig. 2.1: Eggplants

2.1.2. Importance of Eggplants

Eggplants are an essential crop in India, providing several benefits, including:

- **Food security**: Eggplants are a staple vegetable in many Indian cuisines, providing essential nutrients like vitamins, minerals, and antioxidants.
- **Economic benefits:** Eggplant cultivation provides income for millions of farmers, traders, and related industries, contributing significantly to the country's agricultural economy.
- **Cultural significance**: Eggplants have cultural and traditional significance in India, featuring in various cuisines, particularly in North Indian, South Indian, and Bengali cooking.
- **Different Forms**: Eggplant varieties in India include the popular ,Brinjal' or ,Baingan', which comes in various shapes, sizes, and colors, including purple, green, and white.

2.1.3. Agroclimatic Conditions Suitable for Growing

Eggplants can be cultivated round the year in regions with mild winter. For eggplant cultivation, a mild winter is typically characterized by:

- **Temperature**: Average temperatures ranging from 10-20°C (50-68°F) during the winter months (December-February).
- **Minimum temperature**: Not falling below 8-10°C (46-50°F) for an extended period.
- **Frost-free period**: A minimum of 150-200 frost-free days to allow for uninterrupted growth and fruiting.

  In India, regions with mild winters suitable for eggplant cultivation typically experience:
- **Average winter temperatures**: 12-18°C (54-64°F) in the plains and 8-12°C (46-54°F) in the hills.
- **Lowest temperature**: Rarely falls below 5-6°C (41-43°F) in the plains and 2-3°C (36-37°F) in the hills.

Brinjal can be grown on all kind of soils; generally well drained silt loam and clay loam soils are preferred. It is a warm season crop and susceptible to severe frost. Climatic conditions especially low temperature during the cool season is desirable for successful brinjal production.

2.1.4. Optimum period for raising nursery and transplanting

The optimum period for nursery and planting of eggplants varies depending on the region, climate, and weather conditions. However, here are some general guidelines:

- The nursery period for eggplants typically begins 6-8 weeks before the last frost date in temperate climates. In tropical and subtropical climates, seeds can be sown in the nursery throughout the year, except during extreme weather conditions. It's essential to maintain a nursery temperature of 24-27°C (75-81°F) during the day and above 18°C (64°F) at night.
- Seedlings are usually ready for transplanting when they have 2-3 sets of leaves, typically 4-5 weeks after sowing. Before transplanting, it's crucial to harden off seedlings by gradually exposing them to outdoor conditions over 7-10 days.
- In temperate climates, eggplants are typically planted in the field when the soil has warmed up to at least 15°C (59°F), usually 1-2 weeks after the last frost date. In tropical and subtropical climates, eggplants can be planted throughout the year, except during extreme weather conditions. The ideal soil for eggplants is well-draining with a pH between 6.0 & 6.5.

- **In specific regions, the planting times for eggplants are**: In India, the ideal time to plant eggplants is from February to April or September to November. In the United States, eggplants are typically planted in late spring to early summer, when the soil has warmed up. In Southeast Asia, eggplants can be planted throughout the year, except during the hottest months.

2.1.5. Optimum weather & agroclimatic conditions for high productivity

Eggplants are warm-season crops that thrive in favorable weather conditions. Here are the ideal conditions for high productivity of eggplants:

- Eggplants require a long growing season with warm temperatures. The optimal temperature for growth is between 25-30°C (77-86°F). Temperatures above 35°C (95°F) can lead to flower drop and reduced fruit set.
- Adequate moisture is essential for eggplant growth. The crop requires consistent soil moisture, especially during flowering and fruiting stages. Aim for about 600-800 mm of rainfall or irrigation per growing season.
- Eggplants prefer full sun to produce well, requiring at least 6 hours of direct sunlight per day.
- Soil with good drainage and a pH range of 6.0-6.5 is ideal for eggplant cultivation. The crop is sensitive to waterlogged soil, which can lead to root rot and other issues.
- A relative humidity of 60-80% is suitable for eggplant growth. High humidity can lead to disease issues, while low humidity can cause blossom drop.
- Eggplants are susceptible to wind damage, which can reduce fruit quality and yield. Providing wind protection, such as using row covers or planting in a sheltered location, can help minimize damage.

2.1.6. Effect of Extreme Weather Events

Extreme weather events can significantly impact eggplant cultivation, leading to reduced yields, lower quality produce, and economic losses for farmers. Here's how different extreme weather events affect eggplant cultivation:

- **Drought:** Drought stress can cause blossom drop, reducing fruit set and yield. Prolonged drought can lead to plant death, resulting in significant yield losses.
- **Flooding**: Excessive rainfall or flooding can cause root rot, leading to plant death. Flooded conditions also favor disease development, such as fungal and bacterial infections.

- **Heatwaves**: Prolonged high temperatures (above 35°C/95°F) can cause flower drop, reducing fruit set and yield. Heat stress can also lead to sunscald, resulting in damaged fruit.
- **Hailstorms**: Hail damage can cause physical injury to plants, leading to reduced yields and lower quality produce.
- **Windstorms**: Strong winds can cause physical damage to plants, reducing yields and making them more susceptible to disease and pests.

2.2. Okra (*Abelmoschus esculentus*)

Okra also known as lady's finger or bhindi, is a popular vegetable crop grown in many parts of India (Fig. 2.2).

Fig. 2.2: Okra crop

Here's an overview of the area and production of okra, regions grown, and its importance in the Indian context:

2.2.1. Area and Production

Okra is grown in almost all states of India, with the majority of production coming from the states of Maharashtra, Gujarat, Uttar Pradesh, Bihar, and West Bengal. According to the Directorate of Economics and Statistics, Department of Agriculture, Cooperation and Farmers Welfare, Government of India, the total production of okra in India was approximately 6.3 million metric tons in 2020-21.

2.2.2. Regions Grown

Okra is widely grown in various regions of India, including the Indo-Gangetic Plain, the Deccan Plateau, and the Eastern Coastal Plains. The crop is grown in both in *kharif* (summer) and *rabi* (winter) seasons, with the kharif season accounting for the majority of production.

2.2.3. Importance

Okra is an important crop in India for several reasons. Firstly, it is a rich source of nutrients, including vitamins, minerals, and antioxidants. Okra is also high in dietary fiber, making it a good food for digestive health. In addition to its nutritional value, okra is also an important crop for food security, particularly in rural areas where it is a staple vegetable.

- Okra is a high-value crop that can be grown on a small scale, making it an attractive option for small-scale farmers. Additionally, okra is a versatile crop that can be consumed in various forms, including fresh, frozen, and dried, making it a valuable crop for both domestic consumption and export.
- Furthermore, okra has also been recognized for its potential health benefits, including its ability to help manage diabetes, reduce inflammation, and improve cardiovascular health. Okra has also been used in traditional Indian medicine, such as Ayurveda and Unani, for various purposes, including treating fever, rheumatism, and digestive problems.
- In India, okra is also an important crop for cultural and social reasons. It is a popular ingredient in various Indian cuisines, including North Indian, South Indian, and East Indian. Okra is also used in various traditional Indian dishes, such as sambar, dal, and curries.
- Overall, okra is an important crop in India that plays a significant role in food security, economic development, health benefits, and cultural significance. Okra is a warm-season crop that thrives in tropical and subtropical regions of India.

2.2.4. Weather and Agroclimatic Conditions Suitable for Growing Okra

The ideal weather and agroclimatic conditions for growing okra in India are as follows:

- **Temperature**: Okra requires a warm temperature to grow, with an optimal temperature range of 20-35°C. The crop is sensitive to frost and cannot tolerate temperatures below 10°C. The ideal temperature for seed germination is between 25-30°C, while the optimal temperature for fruiting is between 20-25°C.
- **Moisture**: Okra requires adequate moisture, especially during the fruiting stage. The crop requires a well-distributed annual rainfall of at least 600-800 mm. However, excessive rainfall can lead to fungal diseases and reduce fruit quality. The ideal relative humidity for okra

growth is between 60-80%.

- **Soil:** Okra can grow on a wide range of soils, but it prefers well-drained, fertile soils with a pH range of 6.0-7.0. The crop is sensitive to waterlogging and cannot tolerate saline or alkaline soils.
- **Sunlight**: Okra requires full sunlight to grow, with a minimum of 6 hours of direct sunlight per day. The crop responds well to bright, sunny conditions, and shaded conditions can lead to reduced fruiting.
- **Wind**: Okra is sensitive to strong winds, which can damage the plants and reduce fruit quality. The crop prefers calm or gentle breezes, especially during the fruiting stage.
- **Agroclimatic Regions**: Okra can be grown in various agroclimatic regions of India, including the Indo-Gangetic Plain, the Deccan Plateau, and the Eastern Coastal Plains. The crop is widely grown in the states of Maharashtra, Gujarat, Uttar Pradesh, Bihar, and West Bengal.
- **Seasonality**: Okra is typically grown during the kharif (summer) season in India, which coincides with the southwest monsoon. The crop is sown in June-July and harvested in September-October. In some regions, okra is also grown during the rabi (winter) season, which coincides with the northeast monsoon.

Overall, okra requires a warm and humid climate with adequate moisture and sunlight to grow. The crop is sensitive to extreme weather conditions, such as frost, drought, and excessive rainfall, and requires well-drained, fertile soils to thrive.

2.2.5. Optimum weather conditions for high productivity of okra

Following weather elements are requied for highr productivity of okra.

- **Temperature:** It is one of the most critical factors affecting okra productivity. The ideal temperature for okra growth and development is between 20-30°C. Temperatures above 35°C can lead to flower drop, reduced fruit set, and lower yields. On the other hand, temperatures below 15°C can slow down okra growth and reduce yields.
- **Relative humidity**: It also plays a crucial role in okra productivity. A relative humidity of 60-80% is ideal for okra growth and development. High humidity can lead to fungal diseases, while low humidity can cause moisture stress.
- **Rainfall**: Okra requires adequate rainfall or irrigation for optimal growth and development. The crop requires a well-distributed annual rainfall of at least 600-800 mm. However, excessive rainfall can lead to waterlogging, which can reduce okra yields.

- **Sunlight**: It is essential for okra growth and development. The crop requires full sunlight to produce high yields. Okra plants grown in shaded conditions can produce lower yields and may be more susceptible to diseases.
- **Wind**: Wind protection is also important for okra productivity. Okra plants are sensitive to strong winds, which can cause physical damage and reduce growth. Providing wind protection, such as windbreaks or row covers, can help to reduce wind damage and promote healthy growth.

In terms of weather conditions, okra is typically grown during the kharif (summer) season in India, which coincides with the southwest monsoon. The crop is sown in June-July and harvested in September-October. In some regions, okra is also grown during the rabi (winter) season, which coincides with the northeast monsoon. Overall, the optimum weather conditions for high productivity of okra in India include temperatures between 20-30°C, relative humidity of 60-80%, adequate rainfall or irrigation, full sunlight, and wind protection.

2.2.6. Effect of Extreme Weather Extreme Weather Conditions on Okra

Extreme weather conditions can have a significant impact on okra crops, leading to reduced yields, lower quality produce, and even complete crop failure.

- Drought can lead to moisture stress, causing okra plants to become stunted and leading to reduced yields. It can also cause okra flowers to drop, reducing fruit set. On the other hand, excessive rainfall can lead to waterlogging, causing okra roots to rot and leading to plant death. Excessive rainfall can also lead to nutrient leaching, reducing soil fertility.
- High temperatures above 35°C can cause okra flowers to drop, reducing fruit set. High temperatures can also lead to sunscald, causing okra fruits to become discolored. Low temperatures below 15°C can slow down okra growth, leading to reduced yields. Low temperatures can also make okra plants more susceptible to diseases.
- Floods can cause okra plants to become waterlogged, leading to root rot and plant death. Floods can also lead to soil erosion, reducing soil fertility. Heat waves can cause okra plants to become stressed, leading to reduced yields. Heat waves can also cause okra fruits to become discolored.

- Cold waves can cause okra plants to become damaged, leading to reduced yields. Cold waves can also make okra plants more susceptible to diseases. Windstorms can cause okra plants to become damaged, leading to reduced yields. Windstorms can also cause okra fruits to become discolored.
- Hailstorms can also cause significant damage to okra crops, leading to reduced yields and lower quality produce. To mitigate the effects of extreme weather conditions on okra crops, farmers can use various strategies such as irrigation, mulching, crop insurance, and planting drought-tolerant or disease-resistant varieties.

2.3. Tomatoes (*Solanum lycopersicum*)

Tomatoes are one of the most widely cultivated and consumed vegetables in India (Fig. 2.3).

Fig. 2.3: Tomato crop

Here are some key facts about tomato cultivation in India:

- **Area**: Tomatoes are grown in almost all states in India, with the largest producers being Maharashtra, Karnataka, Uttar Pradesh, Bihar, and West Bengal. As per the 2019-2020 data, the total area under tomato cultivation in India was approximately 7.8 lakh hectares.
- **Production**: India is the second-largest producer of tomatoes in the world, accounting for around 22% of global production. In 2019-2020, India produced approximately 22.2 million metric tons of tomatoes.
- **Importance**: Tomatoes are a vital crop in India, contributing significantly to the country's food security and economy. They are a rich source of essential nutrients like vitamin C, potassium, and lycopene, making them a staple in Indian cuisine. Tomatoes are used in a variety of dishes, from curries and sauces to salads and juices. The crop also provides livelihood support to millions of farmers and agricultural laborers across the country.

2.3.1. Weather and Climate Requirements of Tomatoes

In India, the weather and agroclimatic conditions suitable for growing tomatoes which is a warm-season crop that thrives well in temperate climates also with adequate moisture are:

- The optimum temperature for tomato growth is between 20-25°C during the day and 15-20°C at night. Tomatoes can tolerate temperatures between 10-35°C, but extreme temperatures below 10°C or above 35°C can cause damage or reduce yields.
- Adequate rainfall of 600-800 mm per annum, well-distributed throughout the growing season, is necessary for tomato growth. Supplemental irrigation is necessary during dry spells or in areas with low rainfall.
- Tomatoes prefer well-draining soil, such as sandy loam or loamy soils with a pH range of 6.0-7.0. Waterlogged soils should be avoided, as they can lead to root rot and other diseases.
- In India, the tropical regions of Maharashtra, Karnataka, Tamil Nadu, and Kerala are suitable for tomato cultivation. The subtropical regions of Uttar Pradesh, Bihar, and West Bengal, as well as the temperate regions of Himachal Pradesh, Jammu and Kashmir, and Uttarakhand, are also suitable.
- **The best seasons for growing tomatoes in India are**
  - Kharif season (June-September) for tropical and subtropical regions
  - Rabi season (October-February) for subtropical and temperate regions
  - Summer season (March-May) for temperate regions and for growing cherry tomatoes and other varieties that mature quickly.

2.3.2. Optimum Weather Conditions Suitable for Raising Nursery and Transplantation of Tomoto

For successful tomato cultivation in India, it‘s crucial to provide optimum weather conditions during the nursery and transplantation stages.

- During the nursery stage, the ideal temperature is between 20-25°C during the day and 15-20°C at night. The relative humidity should be maintained between 60-80%, with partial shade, especially during peak summer months. The soil temperature should be around 20-25°C for optimal seed germination.
- For transplantation, the ideal temperature is between 22-28°C during the day and 18-22°C at night. The relative humidity should be maintained between 50-70%, with full sun, but avoiding intense heat during peak

summer months. Consistent soil moisture levels should be maintained, avoiding waterlogging.

- In tropical regions, seedlings should be transplanted 7-10 days after sowing, during cloudy weather or early morning hours. In subtropical and temperate regions, transplantation should occur 10-14 days after sowing, during mild weather with no risk of frost.
- It's essential to avoid transplanting during extreme weather conditions like heavy rainfall, intense heat, or frost. Monitoring weather forecasts can help plan transplantation accordingly, minimizing exposure to adverse weather conditions.

2.3.3. Weather and Agroclimatic Conditions Favorable for High Productivity of Tomoto in India

Tomato is one of the most widely cultivated and consumed vegetables in India. To achieve high productivity of tomatoes, it is essential to provide favorable weather and agroclimatic conditions. The ideal conditions for high tomato productivity in India are discussed below:

- Temperature is a critical factor influencing tomato productivity. The optimum temperature for tomato growth and development is between 20-25°C during the day and 15-20°C at night. Temperatures above 35°C or below 10°C can significantly reduce tomato yields. In India, the temperature requirements for tomato cultivation vary across different regions. In tropical regions, tomatoes are grown during the winter months when the temperature is relatively cooler. In subtropical and temperate regions, tomatoes are grown during the summer months when the temperature is warmer.
- Relative humidity also plays a vital role in determining tomato productivity. A relative humidity of 60-80% is ideal for tomato growth. High humidity can lead to fungal diseases, while low humidity can cause moisture stress.
- Light is another essential factor influencing tomato productivity. Tomatoes require full sun to produce well, but excessive sunlight can cause sunscald and reduce yields. In India, the duration and intensity of sunlight vary across different regions. In tropical regions, tomatoes are often grown under partial shade to protect them from intense sunlight.
- Soil temperature is also crucial for tomato growth. The ideal soil temperature for tomato seed germination is between 20-25°C. Soil temperatures above 30°C or below 15°C can significantly reduce seed germination and seedling emergence.

- In addition to temperature, relative humidity, light, and soil temperature, wind speed and direction also influence tomato productivity. Strong winds can damage tomato plants, reduce yields, and increase the risk of disease transmission.
- In India, the agroclimatic conditions favorable for high tomato productivity vary across different regions. The tropical regions of Maharashtra, Karnataka, Tamil Nadu, and Kerala are suitable for tomato cultivation during the winter months. The subtropical regions of Uttar Pradesh, Bihar, and West Bengal are suitable for tomato cultivation during the summer months. The temperate regions of Himachal Pradesh, Jammu and Kashmir, and Uttarakhand are suitable for tomato cultivation during the summer months.

Overall, the weather and agroclimatic conditions favorable for high tomato productivity in India include temperatures between 20-25°C, relative humidity of 60-80%, full sun to partial shade, soil temperature between 20-25°C, and moderate wind speed. By providing these favorable conditions, farmers can achieve high tomato productivity and improve their livelihoods.

2.3.4. Effect of Extreme Weather Events on Tomato Crop

Extreme weather events, such as droughts, floods, heatwaves, and cold snaps, can significantly impact tomato crops, leading to reduced yields, lower quality produce, and economic losses for farmers.

- Drought can cause water stress in tomato plants, leading to reduced growth, flower drop, and lower fruit set. Drought-stressed tomato plants are more vulnerable to pests and diseases, such as nematodes, whiteflies, and fungal infections. This can result in reduced tomato yields, as plants may not receive enough water to produce fruit.
- Floods can cause root rot and other diseases in tomato plants, such as bacterial canker and fungal infections. Flood water can also leach nutrients from the soil, leading to nutrient deficiencies in tomato plants. This can result in reduced tomato yield, as plants may be damaged or killed by excess water.
- Heatwaves can cause heat stress in tomato plants, leading to reduced growth, flower drop, and lower fruit set. Heatwaves can also cause sunscald, a condition where the fruit becomes damaged due to intense heat. This can result in reduced tomato yields, as plants may not be able to tolerate the high temperatures.
- Cold snaps can cause chilling injury in tomato plants, leading to reduced growth, flower drop, and lower fruit set. Cold snaps can also cause frost

damage, which can kill tomato plants or damage fruit. This can result in reduced tomato yields, as plants may be damaged or killed by cold temperatures.

- To mitigate the effects of extreme weather events, farmers can implement efficient irrigation systems and management practices, purchase crop insurance, and adopt agroforestry practices, such as planting trees or using mulch. Breeding tomato varieties that are resilient to extreme weather events can also help farmers adapt to climate change.

2.4. Chillies (*Capsicum annuum*)

Chillies are a vital crop in India, playing a significant role in the country's economy, cuisine, and culture (Fig. 2.4).

Fig. 2.4: Chillies crop

The overview of chilli cultivation is:

- **Area and Production:** India is the world's largest producer, consumer, and exporter of chillies, accounting for over 60% of global production. According to the Indian Spice Board, India produces over 1.5 million tonnes of chillies annually, with the majority coming from the states of Andhra Pradesh, Telangana, Karnataka, and Maharashtra.
- **Regions Grown:** Chillies are grown in various regions across India, including:
  - Andhra Pradesh: Known for its famous Guntur chillies, which are renowned for their intense heat and flavor.
  - Telangana: The state is a significant producer of chillies, with the majority coming from the districts of Warangal and Karimnagar.
  - Karnataka: The state is known for its Byadgi chillies, which are prized for their mild flavor and vibrant red color.
  - Maharashtra: The state is a significant producer of chillies, with the majority coming from the districts of Ahmednagar and Jalgaon.

- **Importance:** Chillies are an integral part of Indian cuisine, adding flavor, heat, and texture to various dishes. They are also a vital crop for many farmers, providing a source of income and employment. Additionally, chillies have several health benefits, including antioxidant and anti-inflammatory properties.
- **Export Potential:** India's chilli export industry has significant potential, with the country exporting over 50,000 tonnes of chillies annually. The major export destinations for Indian chillies include:
  - China
  - Southeast Asia
  - The Middle East
  - Europe
  - The United States

The export of chillies earns significant foreign exchange for the country, making it an important contributor to India's economy. Chillies are a warm-season crop that thrives in India's tropical climate.

2.4.1. Weather and Agroclimatic Requirements for Growing Chillies in India

Here are the weather and agroclimatic requirements for growing chillies in India:

- **Temperature**: Chillies require a warm temperature to grow, with optimal temperatures ranging from 20°C to 30°C. Temperatures above 35°C can lead to flower drop and reduced fruit set.
- **Rainfall**: Chillies require adequate moisture, especially during the fruiting stage. Well-distributed rainfall of around 600-800 mm is ideal for chilli cultivation. However, excessive rainfall can lead to fungal diseases and root rot.
- **Soil:** Chillies can grow in a wide range of soils, but well-draining, fertile soils with a pH range of 6.0-7.0 are ideal.
- **Agroclimatic Regions:** Chillies can be grown in various agroclimatic regions in India, including:
  - Tropical regions: Andhra Pradesh, Telangana, Karnataka, and Tamil Nadu.
  - Subtropical regions: Maharashtra, Gujarat, and parts of Madhya Pradesh.
  - Semi-arid regions: Rajasthan, Haryana, and parts of Punjab.

- **Sowing Time:** The ideal sowing time for chillies varies depending on the region:
    - In tropical regions, chillies are sown in June-July or September-October.
    - In subtropical regions, chillies are sown in February-March or September-October.
    - In semi-arid regions, chillies are sown in September-October.

2.4.2. Optimum Weather Conditions Favorable for Growing Chillies in India

For raising nursery and transplantation of chillies in India, the following optimum weather conditions are required:

- **Raising Nursery:** Chilli seedlings require a temperature of 25-30°C during the day and above 15°C at night. A relative humidity of 60-80% and clear skies with moderate sunshine are also essential.
- Transplantation : For transplantation, a temperature of 20-25°C during the day and above 10°C at night is ideal. Cloudy or partially cloudy skies with moderate temperatures and adequate soil moisture are also necessary.
- **Ideal Time for Transplantation**: In tropical regions, transplant chilli seedlings 30-40 days after sowing, when the weather is cloudy or partially cloudy. In subtropical regions, transplant chilli seedlings 35-45 days after sowing, when the temperature is moderate and there is adequate soil moisture.
- Transplantation should be avoided when heavy rainfall or high temperatures are likely to occur. For high productivity of chillies in India, the following conditions are most favorable:
- **Climate:** Chillies require a warm and dry climate with moderate temperatures ranging from 20-30°C. They are sensitive to extreme temperatures, humidity, and rainfall.
- Soil: Well-draining, fertile soils with a pH range of 6.0-7.0 are ideal for chilli cultivation. Soils with good organic matter content and adequate nutrient supply promote healthy plant growth and high yields.
- **Water Management** : Adequate moisture supply, especially during the fruiting stage, is crucial for high yields. However, excessive rainfall or waterlogging can lead to reduced yields and increased disease susceptibility.

- **Crop Management Practices**: Adopting best management practices such as raised bed cultivation, mulching, and integrated nutrient management can significantly enhance chilli productivity. Additionally, timely planting, pests and diseases management, and support for plants through staking or trellising, can also contribute to higher yields.

2.4.3. Weather Conditions Suitable for Ripening of Red Chillies, Drying and Storage

For optimal ripening, harvesting, drying, and storage of chillies, the following weather conditions are favorable:

- **Ripening**: A warm and dry climate with temperatures ranging from 20-30°C and low humidity facilitates uniform ripening of chillies.
- **Harvesting**: Clear skies with moderate temperatures and low humidity make harvesting easier and reduce the risk of fungal diseases.
- **Drying**: A dry climate with low humidity and moderate temperatures (20-30°C) is ideal for drying chillies. Direct sunlight and good air circulation also facilitate efficient drying.
- **Storage**: A cool, dry, and well-ventilated storage area with temperatures ranging from 10-20°C and low humidity helps preserve the quality and shelf life of dried chillies.

2.4.4. Effect of extreme weather events on chillies

Extreme weather events can significantly impact chilli crops, affecting their growth, yield, and quality. Here‘s a detailed overview of the effects:

- **Temperature Extremes:** High temperatures above 35°C can cause flower drop, reduced fruit set, and sunscald, leading to significant yield losses. On the other hand, temperatures below 10°C can damage or kill chilli plants.
- **Drought**: Drought stress can severely impact chilli yields, particularly during the fruiting stage. Drought can lead to reduced plant growth, flower drop, and increased susceptibility to pests and diseases.
- **Floods and Waterlogging:** Excessive rainfall or waterlogging can cause root rot, reduce plant growth, and increase the risk of fungal diseases. Prolonged waterlogging can also lead to nutrient deficiencies and reduced yields.
- **Hailstorms:** Hailstorms can cause significant damage to chilli plants, leading to reduced yields and lower fruit quality. Hail can damage leaves, break stems, and bruise fruits.

- **Windstorms**: Strong winds can damage chilli plants, reduce yield, and increase the risk of disease transmission. Wind can also cause soil erosion, reduce soil moisture, and increase the risk of pest infestations.
- **Climate Change Impacts:** Climate change is projected to increase the frequency and severity of extreme weather events, posing significant challenges to chilli production. Rising temperatures, changing precipitation patterns, and increased frequency of extreme weather events can reduce chilli yields, affect fruit quality, and increase the risk of pests and diseases.

2.5. Bell Pepper (*Capsicum annuum*)

Bell pepper, also known as sweet pepper or capsicum, is a popular vegetable crop in India (Fig. 2.5).

Fig. 2.5: Bell pepper crop

Here's an overview of its cultivation and importance in the country:

- **Area and Production**: India is one of the largest producers of bell peppers in the world. According to the Indian Horticultural Database, bell pepper is grown in an area of approximately 140,000 hectares, with a total production of around 1.8 million metric tons.
- **Growing Regions**: Bell peppers are grown in various regions across India, primarily in Maharashtra, Karnataka, Gujarat, Tamil Nadu, and Andhra Pradesh. These states provide favorable climatic conditions for bell pepper cultivation.
- **Importance**: Bell peppers hold significant importance in India due to their high nutritional value, being rich in vitamins A and C, and antioxidants. They are also versatile in culinary uses, ranging from salads to stir-fries. Additionally, bell peppers have commercial demand, both domestically and internationally, and offer potential for value-added products, such as pickled peppers and pepper powder.

2.5.1. Weather and Agriclimatic Conditions Suitable For Growing Bell Pepper

Bell pepper is a warm-season crop that thrives in India's diverse climate. Here are the suitable weather and agroclimatic conditions:

- **Temperature:** Bell pepper requires a moderate temperature, ranging from 18°C to 30°C, for optimal growth. Temperatures above 35°C can lead to sun scorching, while temperatures below 15°C can cause chilling injury.
- **Rainfall and Moisture**: Bell pepper needs adequate moisture, especially during the fruiting stage. An annual rainfall of 600-800 mm is ideal. However, excessive rainfall can lead to fungal diseases.
- **Soil and Climate Regions**: Bell pepper can be grown in various soil types, but well-draining sandy loam soils are preferred. In India, bell pepper is grown in regions with mild winters and warm summers, such as Maharashtra, Karnataka, and Gujarat, coastal regions with moderate temperatures and high humidity, such as Tamil Nadu and Andhra Pradesh
- **Growing Seasons**: In India, bell pepper is typically grown during two seasons:
    - Kharif season (June-September): Suitable for regions with moderate temperatures and adequate rainfall.
    - Rabi season (October-February): Ideal for regions with mild winters and low rainfall.

2.5.2. Optimum Weather Conditions for Raising Nursery and Transplanting of Bell Pepper

Bell pepper nursery requires specific weather conditions to ensure healthy seedlings. Here are the optimum weather conditions:

- **Nursery Stage**: For raising bell pepper nursery, the ideal weather conditions are:
    - A temperature range of 20-25°C is optimal for seed germination and seedling growth. Temperatures above 30°C can lead to seedling scorching.
    - Relative humidity of 60-80% helps maintain adequate moisture levels, promoting healthy seedling development.
- **Transplanting Stage**: For transplanting bell pepper seedlings, consider the following weather conditions:

- A temperature range of 22-28°C is suitable for transplanting. Avoid transplanting during extreme temperatures.
- Low to moderate wind speeds (less than 10 km/h) help prevent seedling damage during transplanting.
- Adequate soil moisture is essential for transplanting. Ensure the soil is moist but not waterlogged.

- **Ideal time**: The ideal time for transplanting bell pepper seedlings varies by region:
  - In northern India, transplanting is typically done in late March to early April, when the temperature is mild.
  - In southern India, transplanting is usually done in late September to early October, when the temperature is relatively cooler.
  - Avoid transplanting during periods of heavy rainfall, as this can lead to seedling rot and poor establishment.

2.5.3. Optimum Weather Conditions Suitable for High Productivity of Bell Pepper

Bell pepper is a warm-season crop that requires specific weather conditions to achieve high productivity.

- **Temperature:** The ideal temperature for bell pepper growth is between 25-30°C during the day and 18-22°C at night. Temperatures above 35°C can cause sun scorching and reduce fruit set, while temperatures below 15°C can cause chilling injury and reduce plant growth.
- **Relative Humidity**: A relative humidity of 60-80% is ideal for bell pepper growth and fruit production. High humidity can lead to fungal diseases, while low humidity can cause water stress.
- **Sunshine**: Bell pepper requires full sun to produce well, with at least 6 hours of direct sunlight per day. Adequate sunlight promotes healthy plant growth, fruit set, and color development.
- **Rainfall and Irrigation**: Adequate moisture is essential for bell pepper growth, with a total water requirement of around 600-800 mm per growing season. Well-distributed rainfall or supplemental irrigation is necessary to prevent water stress and ensure optimal fruit production.
- **Wind:** Moderate wind speeds of less than 15 km/h can help prevent fungal diseases by promoting air circulation. However, strong winds can damage plants and reduce fruit set.

2.5.4. Effect of Extreme Weather Events on Bell Pepper

Bell pepper is a warm-season crop that is sensitive to extreme weather events. These events can significantly impact bell pepper growth, productivity, and quality.

- **High Temperatures**: Prolonged exposure to high temperatures (above 35°C) can cause sun scorching and fruit damage, reducing fruit set and lower yields. High temperatures also increase water loss and drought stress.
- **Low Temperatures:** Exposure to low temperatures (below 15°C) can cause chilling injury and reduced plant growth, reducing fruit set and lower yields. Low temperatures also increase susceptibility to diseases.
- **Heavy Rainfall and Flooding**: Heavy rainfall and flooding can cause waterlogged soil and root rot, reducing oxygen availability and plant stress. This also increases the risk of fungal diseases.
- **Drought**: Drought conditions can cause water stress and reduced plant growth, reducing fruit set and lower yields. Drought also increases susceptibility to pests and diseases.
- **Strong Winds:** Strong winds can cause physical damage to plants and fruit, reducing fruit set and lower yields. Strong winds also increase the risk of disease transmission.
- **Hailstorms:** Hailstorms can cause physical damage to plants and fruit, reducing fruit set and lower yields. Hailstorms also increase the risk of disease transmission.

To mitigate the effects of extreme weather events, farmers can use strategies such as using greenhouses or shade nets to regulate temperature and protect from wind and hail. Implementing irrigation management systems can optimize water use. Practicing crop insurance and diversification can reduce risk. Using resistant varieties and integrated pest management techniques can reduce disease susceptibility.

2.6. Avocados (*Persea americana*)

Avocado cultivation in India is gaining popularity, particularly in states with suitable climates like Karnataka, Tamil Nadu, Maharashtra, and Kerala (Fig. 2.6). These regions provide the ideal conditions for avocado trees to thrive, with well-drained soil and temperatures between 20 to 30°C .

Fig. 2.6: Avocado crop

- **Avocado Production in India:** India's avocado production is still in its early stages, with most avocados being imported from countries like Mexico and Kenya. However, with the increasing demand for healthy and nutritious food, avocado farming is becoming a lucrative venture for Indian farmers .
- **Importance of Avocados in India**: Avocados are a nutrient-rich fruit, high in proteins, fats, and various essential vitamins and minerals. They are considered a healthy food option, and their popularity is growing rapidly in India. Avocado cultivation can provide a profitable income source for Indian farmers, while also contributing to the country's nutritional security.
- **Area Under Avocado Cultivation**: Although exact figures on the area under avocado cultivation in India are not available, it is reported that avocado plantations are being established in various states including Karnataka, Tamil Nadu, Maharashtra, and Kerala.
- **Challenges and Opportunities**: Despite the growing demand for avocados, Indian farmers face challenges like limited knowledge on avocado cultivation, inadequate irrigation facilities, and pests and diseases. However, with government initiatives and support from research institutions, avocado farming is expected to become a significant contributor to India's fruit production and economy .

2.6.1. Weather and Agroclimatic Conditions for Growing Avocados

Avocado trees require specific weather and agroclimatic conditions to thrive. Here are the ideal conditions:

- **Temperature**: Avocado trees prefer a mild, subtropical climate with average temperatures ranging from 15°C to 30°C. Temperatures above

35°C can cause sunburn, while temperatures below 10°C can damage the tree.

- **Humidity:** Avocado trees thrive in humid environments with an average relative humidity of 60-80%. High humidity helps maintain healthy growth, while low humidity can lead to water stress.
- Rainfall: Avocado trees require adequate rainfall, with an annual rainfall of at least 1,000 mm. Well-distributed rainfall throughout the year ensures consistent growth and fruit production.
- **Soil:** Avocado trees prefer well-draining, fertile soils with a pH range of 6-7. Soils with poor drainage can lead to root rot, while soils with high pH levels can cause nutrient deficiencies.
- **Sunlight**: Avocado trees require full sun to partial shade, with at least 6 hours of direct sunlight per day. Adequate sunlight promotes healthy growth, fruit production, and color development.
- **Wind**: Avocado trees prefer gentle winds, with wind speeds less than 15 km/h. Strong winds can damage trees, reduce fruit set, and increase disease susceptibility.

In India, the states of Karnataka, Tamil Nadu, Maharashtra, and Kerala provide suitable agroclimatic conditions for growing avocados. These regions offer mild winters, warm summers, and well-distributed rainfall, making them ideal for avocado cultivation.

2.7. Conclusion

India's diverse climate and geography provide a favorable environment for growing a wide range of fruit-type vegetables, including tomatoes, bell peppers, avocados etc. These vegetables are not only rich in nutrients and flavor but also contribute significantly to India's agricultural economy. With the increasing demand for healthy and nutritious food, the cultivation of fruit-type vegetables is expected to grow, providing opportunities for farmers, processors, and exporters. By adopting improved cultivation practices, investing in research and development, and promoting value addition and processing, India can further enhance its position as a major producer and exporter of fruit-type vegetables.

3

Agroclimatology of Creeper Type Vegetables

Agroclimatology plays a vital role in the cultivation and production of creeper vegetables in India. Creeper vegetables, such as cucumber, bitter gourd, and bottle gourd, are an essential part of Indian cuisine and are grown extensively throughout the country. With its diverse climate and geographical conditions, India offers a suitable environment for the growth of a wide range of creeper vegetables.

In terms of area and production, creeper vegetables account for a significant share of the total vegetable production in India. According to the Ministry of Agriculture and Farmers Welfare, Government of India, the area under creeper vegetable cultivation has been increasing steadily over the years, with a corresponding rise in production. This trend is attributed to the growing demand for these vegetables, both domestically and internationally.

Creeper vegetables are not only an important source of nutrition but also contribute significantly to the country's economy. They are a vital source of income for many small and marginal farmers, who cultivate these crops on a large scale. Moreover, the export of creeper vegetables earns significant foreign exchange for the country.

The importance of creeper vegetables in India can be gauged from the fact that they are an integral part of various traditional and cultural festivals. For instance, bitter gourd is considered a auspicious vegetable in many Indian communities and is often prepared during special occasions.

This chapter aims to provide a comprehensive overview of the agroclimatology of creeper vegetables in India, including their area, production, and importance. It will delve into the various factors that affect the growth and productivity of these vegetables, such as temperature, rainfall, and soil type. Additionally, it will discuss the various initiatives and strategies adopted by the government and other stakeholders to promote the cultivation and production of creeper vegetables in the country.

3.1. Growth Habits of Creeper Vegetables

In India, creeper vegetables are a common sight in many gardens and farms. These vegetables are characterized by their trailing or climbing habit, which requires support for optimal growth. However, some creeper vegetables can be spread on the ground, eliminating the need for support. Here, we'll discuss both types of creeper vegetables commonly found in India (Fig. 3.1).

Fig. 3.1: Creeper vegetable crops

3.1.1. Creeper Vegetables Requiring Support

Several popular creeper vegetables in India require support to grow upright and produce a bountiful harvest. Some examples include

- **Cucumber (Kheera)**: Cucumbers are one of the most widely cultivated creeper vegetables in India. They require a trellis or a fence to climb on, which helps to keep the fruit off the ground and promotes better air circulation.
- **Bitter Gourd (Karela)** : Bitter gourd is another popular creeper vegetable in India that requires support. It can be trained to climb up a trellis or a fence, which helps to keep the fruit clean and promotes better growth.
- **Bottle Gourd (Lauki)**: Bottle gourd is a versatile creeper vegetable that requires support to grow. It can be trained to climb up a trellis or a fence, which helps to keep the fruit off the ground and promotes better air circulation.

- **Snake Gourd (Chichinda)**: Snake gourd is a type of creeper vegetable that requires support to grow. It can be trained to climb up a trellis or a fence, which helps to keep the fruit clean and promotes better growth.

3.1.2. Creeper Vegetables that can be Spread on the Ground

Some creeper vegetables in India can be spread on the ground, eliminating the need for support. Some examples include:

- **Pumpkin (Kaddu)**: Pumpkin is a type of creeper vegetable that can be spread on the ground. It has a sprawling habit and can be grown without support.
- **Squash (Lal Bhopla)**: Squash is another type of creeper vegetable that can be spread on the ground. It has a sprawling habit and can be grown without support.
- **Round Gourd (Tinda)**: Round gourd is a type of creeper vegetable that can be spread on the ground. It has a sprawling habit and can be grown without support.
- **Ridge Gourd (Torai)** : Ridge gourd is a type of creeper vegetable that can be spread on the ground. It has a sprawling habit and can be grown without support.

In conclusion, while some creeper vegetables in India require support to grow, others can be spread on the ground. By understanding the specific requirements of each type of creeper vegetable, farmers and gardeners can optimize their growth and production.

3.2. Cucumber (*Cucumis sativus*)

India is one of the largest producers of cucumber in the world. The country's diverse climate and geographical conditions make it an ideal place for growing a wide range of cucumber varieties (Fig. 3.2).

Fig. 3.2: Cucumber crop

- **Area and Production**: According to the Ministry of Agriculture and Farmers Welfare, Government of India, the area under cucumber cultivation in India is around 1.3 million hectares. The production of cucumbers in India is approximately 7.3 million metric tons per year. The major cucumber-producing states in India are Maharashtra, Gujarat, Uttar Pradesh, and Punjab.

- **Weather Requirements**: Cucumbers are a warm-season crop and require a long growing season. They thrive in temperatures between 25°C to 35°C. The ideal climate for growing cucumbers is characterized by warm days and cool nights. Cucumbers are also sensitive to frost and cannot tolerate temperatures below 10°C.
- **Types of Cucumbers**: India is home to a diverse range of cucumber varieties, each with its unique characteristics and uses. Some of the most common types of cucumbers grown in India include:
  - Slicing cucumbers are the most widely grown variety in India. They are harvested when immature and are used fresh in salads, sandwiches, and as a snack.
  - Pickling cucumbers are smaller in size and are harvested when mature. They are used for making pickles and are an essential ingredient in Indian cuisine.
  - English cucumbers are a popular variety grown in India. They are longer and slender than slicing cucumbers and are often used in salads and as a snack.
  - Beit Alpha cucumbers are a type of pickling cucumber that is widely grown in India. They are small in size and have a sweet, crunchy texture.
- **Importance in India:** Cucumbers are an essential crop in India, providing livelihoods to thousands of farmers and agricultural workers. They are also an important source of nutrition, being low in calories and rich in vitamins and minerals.
  - In addition to their nutritional value, cucumbers also have several medicinal properties. They are used in traditional Indian medicine to treat a range of ailments, including digestive problems and skin conditions.
  - Cucumbers are also an important ingredient in Indian cuisine, being used in a range of dishes, from salads and sandwiches to pickles and chutneys.

Overall, cucumbers are a vital crop in India, providing economic, nutritional, and medicinal benefits to millions of people.

3.2.1. Vegetative Propagation

Cucumbers can be grown by planting stems, a process known as stem cutting or vegetative propagation. This method is often used in commercial cucumber production and can be successful in home gardens as well.

- **Advantages of Growing Cucumbers from Stem Cuttings:** Growing cucumbers from stem cuttings has several advantages:
  - **Faster production**: Cucumbers grown from stem cuttings can produce fruit within 30-40 days, compared to 60-90 days when grown from seeds.
  - **Increased yields:** Stem-cutting-grown cucumbers can produce more fruit per plant, resulting in higher yields.
  - **Disease resistance:** By using stem cuttings from disease-resistant plants, farmers can reduce the risk of disease transmission.n
  - **Genetic uniformity**: Stem-cutting-grown cucumbers will have the same genetic characteristics as the parent plant, ensuring uniformity in fruit quality and production.
- **How to Grow Cucumbers from Stem Cuttings:** To grow cucumbers from stem cuttings, follow these steps:
  - **Select healthy stem cuttings**: Choose stem cuttings from healthy, disease-free cucumber plants. Cuttings should be 10-15 cm long and have at least two nodes.
  - **Prepare the rooting medium**: Fill a container with a well-draining rooting medium, such as a mix of peat moss and perlite.
  - **Plant the stem cutting**: Plant the stem cutting in the rooting medium, burying it up to the node. Water gently.
  - **Provide optimal conditions**: Keep the soil consistently moist and provide warm temperatures (around 25°C) and high humidity.
  - **Transplant**: Once the stem cutting has developed a robust root system and new growth, transplant it into a larger container or directly into the field.
- **Weather Conditions Required for Stem Planting of Cucumbers in India:** In India, the ideal weather conditions for stem planting of cucumbers are:
  - **Temperature**: The optimal temperature for rooting cucumber stem cuttings is between 24°C to 30°C.
  - **Humidity**: High humidity (around 80-90%) is essential for promoting root growth and preventing water loss.

    - **Light**: Cucumber stem cuttings require bright, indirect light. Direct sunlight can cause overheating and reduce rooting success.
    - **Rainfall:** Consistent moisture is crucial for rooting cucumber stem cuttings. In areas with low rainfall, irrigation may be necessary.
- **In India, the best time for stem planting of cucumbers varies by region**
    - **North India**: The ideal time for stem planting is from February to April or September to November, when the temperature is mild.
    - **South India**: Stem planting can be done year-round, except during the hottest summer months (April to June).
    - **East India**: The best time for stem planting is from March to May or September to November, when the temperature is warm and humid.

3.2.2. Challenges and Limitations

While growing cucumbers from stem cuttings can be successful, there are some challenges and limitations to consider:

- **Rooting difficulties**: Not all cucumber varieties root easily from stem cuttings.
- **Disease transmission**: If the parent plant is infected with a disease, it can be transmitted to the new plants grown from stem cuttings.
- **Limited availability of suitable stem cuttings**: Access to healthy, disease-free stem cuttings can be limited, especially for small-scale farmers or home gardeners.

Overall, growing cucumbers from stem cuttings can be a viable option for farmers and gardeners looking to increase yields and reduce production time. However, it's essential to carefully select healthy stem cuttings, provide optimal growing conditions, and consider the specific weather requirements for stem planting in different regions of India.

3.2.3. Optimum weather conditions for sowing

Cucumbers are a warm-season crop that thrives in India's climate. To achieve optimal growth and yield, it's essential to sow cucumbers under the right weather conditions.

- **Temperature:** The ideal temperature for sowing cucumbers in India is between 25°C to 30°C. Temperatures above 35°C can lead to poor germination, while temperatures below 20°C can slow down growth. The optimal temperature for cucumber seed germination is around 28°C.

- **Moisture**: Cucumbers require adequate moisture, especially during the germination and fruiting stages. The soil should be consistently moist but not waterlogged. Ideal relative humidity for cucumber cultivation is between 60% to 80%.
- **Sunshine**: Cucumbers need full sun to produce well, requiring at least 6 hours of direct sunlight per day. However, in regions with extremely high temperatures, providing some afternoon shade can help prevent scorching.
- **Rainfall:** Cucumbers require about 500-600 mm of rainfall per growing season. However, excessive rainfall can lead to fungal diseases and waterlogged soil. Well-distributed rainfall or supplemental irrigation is essential for optimal growth.
- **Wind:** Cucumbers are sensitive to strong winds, which can damage the plants and reduce yields. A gentle breeze is beneficial, but winds above 15 km/h can be detrimental.
- **Sowing Time:** In India, the ideal time to sow cucumbers varies by region:
  - In the plains, sow cucumbers in late spring to early summer (February-March) or late summer to early autumn (August-September).
  - In the hills, sow cucumbers in spring (March-April) or autumn (September-October).

By considering these weather conditions and sowing cucumbers at the right time, you can optimize growth, yields, and fruit quality.

3.2.4. Structural Requirements for Support if Needed

The framework should have a sturdy frame that can support the weight of the creeper, soil, and any additional features like planters or lights. It should be able to withstand various weather conditions, such as rain, sunlight, and extreme temperatures. If using metal frames, they should be rust-resistant or coated to prevent corrosion.

- **Design Requirements:** The framework should have a creeper-friendly design with a suitable structure for the creeper to climb, such as a trellis or a lattice. It should be adjustable to accommodate different types of creepers and their growth patterns. Additionally, the framework should be visually appealing and blend with the surrounding environment.
- **Material Requirements**: Eco-friendly materials like wood, bamboo, or recycled metal should be used for the framework. The materials should require minimal maintenance, such as cleaning or replacement.

Furthermore, the materials should be resistant to pests and diseases to prevent damage to the creeper.

- **Safety Requirements**: The framework should be stable and secure to prevent it from toppling over. It should have a suitable weight capacity to support the creeper and any additional features. If using electrical components, such as lights, they should be safely installed and meet electrical safety standards.
- **Additional Features**: A built-in irrigation system can be beneficial for maintaining the creeper's health. Integrated planters can provide additional space for plants and enhance the overall aesthetic. Strategically placed lights can highlight the creeper's beauty and provide ambiance.

3.2.5. Weather Conditions Favorable for High Productivity

For high productivity in cucumber cultivation, the following optimum weather conditions are required:

- **Temperature**: A temperature range of 25-30°C is ideal for cucumber growth. Temperatures above 35°C can lead to sunscald, reduced pollination, and lower yields. On the other hand, temperatures below 15°C can slow down growth and reduce fruit set.
- **Moisture**: Adequate moisture is essential for cucumber growth. The ideal relative humidity is between 60-80%. Soil moisture should be maintained at around 70-80% of the field capacity. Consistent moisture levels promote healthy growth, fruiting, and fruit quality.
- **Sunshine**: Cucumbers require full sun to produce well, needing at least 6 hours of direct sunlight per day. However, in regions with extremely high temperatures, providing some afternoon shade can help prevent scorching.
- **Rainfall**: Well-distributed rainfall or supplemental irrigation is essential for optimal growth. Cucumbers require around 500-600 mm of rainfall per growing season. Excessive rainfall can lead to fungal diseases, waterlogged soil, and reduced yields.
- **Wind:** While some wind can be beneficial for pollination, strong winds can damage the plants, reduce yields, and increase the risk of disease. Winds above 15 km/h can be detrimental to cucumber productivity.

3.2.6. Optimum Weather Conditions for Different Growth Stages

- Germination: 25-30°C temperature, 70-80% relative humidity, and consistent moisture.

- Seedling establishment: 20-25°C temperature, 60-70% relative humidity, and moderate moisture.
- Fruiting: 25-30°C temperature, 60-80% relative humidity, and adequate moisture.

By maintaining these optimum weather conditions, cucumber growers can promote healthy growth, maximize yields, and achieve high productivity.

3.2.7. Effect of Extreme Weather Events

Extreme weather events can significantly impact cucumber crops, leading to reduced yields, lower quality produce, and economic losses for farmers.

- **Drought:** Drought can lead to water stress, causing plants to become stunted, and reducing yields. Insufficient water can result in smaller fruit sizes, affecting the overall quality and market value of the produce. Weakened plants are more vulnerable to pests and diseases, which can further exacerbate yield losses.
- **Excessive Rainfall and Flooding:** Excessive rainfall can cause waterlogged soil, leading to root rot and plant death. Reduced oxygen availability in the soil can affect root growth and overall plant health. Excessive moisture can create an ideal environment for fungal diseases to develop, such as powdery mildew and downy mildew.
- **High Temperatures and Heatwaves:** Cucumbers are sensitive to high temperatures, which can cause sunscald, heat stress, and reduced yields. High temperatures can affect pollinator activity, reducing fruit set and yields. Cucumbers require more water during hot weather, which can strain water resources and increase irrigation costs.
- **Hailstorms and Strong Winds**: Hailstorms and strong winds can cause physical damage to plants, reducing yields and affecting fruit quality. Damage to leaves and stems can reduce plant growth, affecting the overall yield and quality of the crop. Physical damage can create entry points for diseases, further exacerbating yield losses.

3.2.8. Mitigation Strategies

To mitigate the effects of extreme weather events, farmers can implement efficient irrigation systems and scheduling. Crop insurance can provide financial protection against yield losses. Diversification, soil conservation, and monitoring and forecasting can also help farmers prepare for and respond to extreme weather events.

3.3. Ridge Gourd (*Luffa acutangula*)

Ridge gourd, is a popular vegetable in India and other parts of Asia (Fig. 3.3). Here are some key facts about ridge gourd:

Fig. 3.3: Ridge gourd crop

- **Area and Production:** Ridge gourd is widely cultivated in India, particularly in the states of Maharashtra, Gujarat, Karnataka, and Tamil Nadu. According to the Indian Ministry of Agriculture and Farmers Welfare, the country produced around 1.3 million metric tons of ridge gourd in 2020-21.
- **Importance:** Ridge gourd is an important vegetable crop in India, providing several benefits. It is rich in vitamins, minerals, and antioxidants, making it a healthy addition to various dishes. The crop is also a significant source of income for many farmers in India, particularly small and marginal farmers. Ridge gourd is a versatile vegetable used in various Indian dishes, such as curries, stews, and stir-fries. It has also been used in traditional medicine for its anti-inflammatory, antioxidant, and antimicrobial properties.
- **Cultivation and Climate:** Ridge gourd is a warm-season crop that thrives in tropical and subtropical climates. It requires a temperature range of 25°C to 35°C. Well-drained, fertile soil with a pH range of 6.0 to 7.0 is ideal for ridge gourd cultivation. The crop also requires adequate rainfall, with an annual rainfall of at least 600 mm.

Overall, ridge gourd is an important crop in India, providing nutritional, economic, and culinary benefits.

3.3.1. Weather Conditions Favorable for Sowing Ridge Gourd in India

Ridge gourd is a warm-season crop that thrives in tropical and subtropical climates. It requires a temperature range of 25°C to 35°C. Well-drained, fertile soil with a pH range of 6.0 to 7.0 is ideal for ridge gourd cultivation. The crop also requires adequate rainfall, with an annual rainfall of at least 600 mm.

3.3.2. Weather Conditions Required for High Production

Ridge gourd is a warm-season crop that requires specific weather conditions to thrive in India. Here are the optimum weather conditions required for high productivity of ridge gourd in India:

- **Temperature**: The optimum temperature for ridge gourd growth is between 25-30°C (77-86°F). The ideal germination temperature is 20-25°C (68-77°F), while the fruit set temperature is between 25-28°C (77-82°F). Temperatures above 35°C (95°F) can lead to flower drop and reduced fruit set.
- **Humidity:** A relative humidity of 60-80% is ideal for ridge gourd growth. High humidity during flowering and fruiting stages helps in better pollination and fruit set.
- **Rainfall:** Ridge gourd requires a well-distributed rainfall of 600-800 mm (24-32 in) during the growing season. Excessive rainfall can lead to waterlogged soil and reduced fruit quality.
- **Sunshine:** Ridge gourd requires full sun to produce well, so it should be planted in an area that receives direct sunlight for at least 6 hours a day.
- **Wind**: Moderate wind can help in better pollination, but strong winds can damage the plants.
- **Region-wise Weather Conditions**: In North India, ridge gourd is grown in the summer season (April-June) when the temperature is between 25-35°C (77-95°F). In South India, it is grown in the summer season (February-April) when the temperature is between 25-30°C (77-86°F). In East India, it is grown in the summer season (March-May) when the temperature is between 25-30°C (77-82°F).

By providing ridge gourd plants with these optimum weather conditions, you can promote healthy growth, high productivity, and better fruit quality in India.

3.3.3. Effects of Extreme Weather Events

Extreme weather events can significantly impact ridge gourd production:

- Rising Temperatures and Changing Precipitation Patterns can lead to reduced yields, lower quality produce, and increased susceptibility to diseases [1].
- Ridge gourd is commonly grown during the summer season, and extreme heat waves can cause damage to crops, leading to reduced yields. On the other hand, excessive rainfall can lead to flooding, which can ruin crops and contaminate water supplies.

- Droughts can also have a devastating impact on ridge gourd production, leading to crop failure and reduced yields. In areas prone to drought, farmers may need to adopt drought-resistant agriculture practices to minimize losses [1].
- Furthermore, extreme weather events can also lead to an increased incidence of diseases such as downy mildew, which can significantly impact ridge gourd yields.

To mitigate the effects of extreme weather events, farmers can adopt advanced technologies such as raised bed cultivation, mulching, and drip irrigation. These practices can help reduce soil erosion, conserve water, and promote healthy plant growth.

Overall, extreme weather events can have a significant impact on ridge gourd production, and farmers need to adopt climate-resilient practices to minimize losses and ensure sustainable production.

3.4. Snake Gourd (*Trichosanthes cucumerina*)

Snake gourd is a popular vegetable crop in India, widely cultivated and consumed across the country (Fig. 3.4). It requires support for spreading.

Fig. 3.4: Snake gourd crop

- **Area and Production:** According to the Indian Ministry of Agriculture and Farmers Welfare, during 2020-21:
  - The total area under snake gourd cultivation in India was approximately 63.5 thousand hectares.
  - The total production of snake gourd in India was around 1,349.8 thousand metric tons.
- **Importance of Snake Gourd in India**: Snake gourd is an important crop in India due to its:

- **Nutritional value**: Rich in vitamins, minerals, and antioxidants, making it a healthy addition to various Indian dishes.
- **Culinary uses**: Used in a variety of Indian recipes, such as curries, stir-fries, and salads.

- **Economic significance**: Provides livelihood opportunities for farmers, traders, and other stakeholders involved in its cultivation, processing, and marketing.
  - **Traditional medicine**: Used in Ayurvedic medicine to treat various health ailments, such as fever, digestive issues, and respiratory problems.
  - **Cultural significance**: Featured in various Indian festivals and ceremonies, symbolizing good health, prosperity, and longevity.

3.4.1. Weather and Climatic Conditions Suitable for Growing Snake Gourd

Snake gourd is a popular vegetable crop in India, widely cultivated and consumed across the country. Here are the ideal weather and climate conditions for growing snake gourd in India:

- **Temperature:** The optimum temperature for growing snake gourd in India is between 25-35°C (77-95°F). The minimum temperature required is 15°C (59°F), while the maximum temperature that the crop can tolerate is 40°C (104°F). Regions with mild winters and hot summers, such as Maharashtra, Gujarat, and Uttar Pradesh, are suitable for snake gourd cultivation.
- **Rainfall:** Snake gourd requires an annual rainfall of 600-1200 mm. Well-distributed rainfall ensures consistent moisture levels, which are essential for healthy plant growth. Regions with high rainfall, such as Kerala, Tamil Nadu, and Karnataka, are ideal for snake gourd cultivation.
- **Humidity**: A relative humidity of 60-80% is ideal for growing snake gourd in India. High humidity promotes healthy plant growth and fruit development.
- **Soil**: Snake gourd prefers well-draining soil, such as sandy loam or clay loam with good drainage. The ideal soil pH is 6.0 to 7.0. Regions with fertile soil, such as the Indo-Gangetic Plain, are suitable for snake gourd cultivation.
- **Sunlight**: Snake gourd requires full sun to grow, needing direct sunlight for at least 6 hours a day.

- **Growing Season**: In India, snake gourd is typically planted in *late spring to early summer*, spring-summer from February to May. The growing period is around 100-120 days from sowing to harvest.

3.4.2. Regions Suitable for Snake Gourd Cultivation

Snake gourd can be grown in various regions across India, including:

- Maharashtra: Known for its high production of snake gourd, Maharashtra is an ideal region for cultivation.
- Gujarat: The state's warm and humid climate makes it suitable for snake gourd cultivationn
- Uttar Pradesh: The Indo-Gangetic Plain in Uttar Pradesh provides fertile soil and a favorable climate for snake gourd cultivation.
- Kerala: The state's high rainfall and humid climate make it an ideal region for snake gourd cultivation.
- Tamilnadu. Warm and humid regions of Tamilnadu are suitable for snake gourd cultivation.

By providing the optimal weather and climate conditions, snake gourd plants can thrive in India, producing high-quality fruits and maximizing yields.

3.4.3. Optimum Weather Conditions Suitable for Planting of Snake Gourd

The ideal time for planting snake gourd is during the spring season, from February to April, and the summer season, from May to June.

- **Temperature**: The temperature ranges from 25-30°C (77-86°F) during the day and 18-22°C (64-72°F) at night, is suitable for seed germination and seedling growth.
- **Rainfall:** Well-distributed rainfall of 100-150 mm per month ensures consistent moisture levels in the soil, promoting healthy plant growth. A relative humidity of 60-80% and high humidity also support seed germination and seedling growth.
- **Sunlight**: Snake gourd requires direct sunlight for 6-8 hours a day, with minimal cloud cover. Low wind speed, less than 15 km/h (9.3 mph), prevents damage to seedlings. The soil temperature should be between 20-25°C (68-77°F) for optimal seed germination, and adequate soil moisture supports seedling growth.

It is essential to avoid planting snake gourd during extreme temperatures, above 35°C (95°F) or below 15°C (59°F), heavy rainfall, and drought conditions. By planting snake gourd during the optimal weather conditions, farmers can ensure healthy plant growth, high yields, and better quality fruits.

3.4.4. Weather and Agroclimatic Conditions Required for High Productivity

Snake gourd is a warm-season crop that thrives in tropical and subtropical climates. To achieve optimal productivity, the following weather and climatic conditions are essential:

- **Temperature**: The ideal air temperature for snake gourd growth ranges from 25-35°C (77-95°F). Temperatures above 35°C (95°F) can lead to flower drop and reduced fruit set, while temperatures below 15°C (59°F) can slow down plant growth. The ideal soil temperature for snake gourd growth ranges from 20-25°C (68-77°F). Adequate soil moisture ensures healthy seed germination and seedling growth.
- **Rainfall**: Well-distributed rainfall of 600-1200 mm (24-47 in) per annum is necessary for optimal snake gourd production. Consistent moisture levels in the soil ensure healthy plant growth and fruit development.
- **Relative Humidity**: A relative humidity of 60-80% is ideal for snake gourd growth. High humidity promotes healthy plant growth and fruit development, while low humidity can lead to pest and disease issues.
- **Sunlight**: Snake gourd requires full sun to grow, needing direct sunlight for at least 6 hours a day. Minimal cloud cover ensures adequate sunlight for photosynthesis and plant growth.
- **Wind**: Low wind speed, less than 15 km/h (9.3 mph), prevents damage to plants and promotes healthy growth.

The ideal regions for snake gourd cultivation are those with warm and humid climates, such as Maharashtra, Gujarat, Uttar Pradesh, Kerala, and Tamil Nadu. By providing the optimal weather and climatic conditions, farmers can achieve high snake gourd productivity and better quality fruits.

3.4.5. Effect of Extreme Weather Events

Snake gourd, like many other crops, is vulnerable to extreme weather events, which can significantly impact its growth, productivity, and quality.

- **Drought**: It can cause water stress in snake gourd plants, leading to reduced growth, lower yields, and poor fruit quality. Drought-stressed snake gourd plants are also more susceptible to pests and diseases.
- **Flood**: It can cause root rot in snake gourd plants, leading to plant death. Excessive water can also lead to nutrient deficiency in the soil, affecting snake gourd plant growth. Additionally, floods can delay harvesting, reducing the market value of the crop.
- **Heatwaves**: These can cause flower drop in snake gourd plants, reducing fruit set. High temperatures can also cause sun scorch on snake gourd

fruits, reducing their quality. Furthermore, snake gourd plants require more water during heatwaves, which can be challenging during periods of drought.

- **Cyclones and strong winds**: They can damage snake gourd plants, reducing yields and affecting fruit quality. Strong winds can also cause fruit loss, reducing the market value of the crop.
- **Extreme rainfall events**: These can cause soil erosion, affecting soil fertility and snake gourd plant growth. Excessive rainfall can also cause waterlogging, leading to root rot and other problems in snake gourd plants.

To mitigate the effects of extreme weather events on snake gourd, farmers can adopt strategies such as diversification, irrigation management, crop insurance, agroforestry practices, and growing climate-resilient varieties. By adopting these strategies, farmers can reduce the impact of extreme weather events on snake gourd and ensure sustainable production.

3.5. Bottle Gourd (*Lagenaria siceraria*)

Bottle gourd is a popular vegetable crop in India, widely cultivated and consumed across India (Fig. 3.5).

Fig. 3.5: Bottle gourd crop

- **Area and Production:** According to the Department of Agriculture, Cooperation and Farmers Welfare, Government of India, the area under bottle gourd cultivation in India is significant. Here are some recent key statistics:
  - Area: 123 thousand hectares (2020-21)
  - Production: 1,834 thousand metric tons (2020-21)
  - Productivity: 14.9 metric tons per hectare (2020-21)

- **Importance of Bottle Gourd in India:** Bottle gourd is an important crop in India due to its nutritional value, versatility in culinary uses, economic significance, cultural significance, and drought tolerance.
  - Bottle gourd is rich in vitamins, minerals, and antioxidants, making it a healthy addition to various dishes. It is used in a variety of preparations, such as curries, soups, salads, and snacks.
  - The crop provides livelihood opportunities for farmers, traders, and related industries. It is also featured in traditional Indian medicine, Ayurveda, and used in various cultural and religious ceremonies. Its ability to thrive in water-scarce conditions makes it an attractive option for farmers in drought-prone areas.

Bottle gourd is a climbing plant that requires support to grow and produce fruit efficiently. The plant's vines can spread quite long, typically ranging from 3 to 5 meters (10 to 16 feet), and need a trellis or other support system to keep them upright and promote healthy growth.

3.5.1. Types of Supports

- **Trellis**: A trellis is a common support system used for bottle gourd. It can be made of wood, metal, or plastic and provides a structure for the vines to climb.
- **Arbor:** An arbor is a more elaborate support system that provides shade and a structure for the vines to climb.
- **Fence**: A fence can also be used as a support system for bottle gourd. The vines can be trained to climb along the fence.

3.5.2. Benefits of Providing Support

- **Improved Fruit Production**: Providing support helps to keep the fruit off the ground, reducing rot and disease.
- **Reduced Soil Moisture Loss**: By keeping the vines off the ground, support systems help to reduce soil moisture loss through evaporation.
- **Easier Harvesting**: Supporting the vines makes it easier to harvest the fruit, as it is more accessible and visible.

Overall, providing support for bottle gourd is essential for promoting healthy growth, improving fruit production, and reducing maintenance.

3.5.3. Weather Conditions Suitable for Planting Bottle Gourd

Bottle gourd is a warm-season crop that thrives in temperate climates with adequate moisture. To ensure optimal growth and fruit production, it's essential to plant bottle gourd in suitable weather conditions.

- **Temperature**: Bottle gourd prefers a warm temperature, typically above 18°C (64°F). The ideal temperature range for germination is between 25°C (77°F) and 30°C (86°F). Temperatures above 35°C (95°F) can lead to scorching, while temperatures below 15°C (59°F) can slow down growth.
- **Moisture**: Adequate moisture is crucial for bottle gourd growth. The crop requires consistent soil moisture, especially during the germination and fruiting stages. However, excessive rainfall or waterlogging can lead to root rot and other diseases.
- **Humidity**: Bottle gourd prefers a relatively high humidity, typically above 50%. However, high humidity can also lead to fungal diseases. A balanced humidity level, combined with good air circulation, is essential for healthy growth.
- **Sunlight**: Bottle gourd requires full sun to produce well, typically requiring at least 6 hours of direct sunlight per day. However, in regions with intense sunlight, providing some shade during the hottest part of the day can help prevent scorching.
- **Wind**: While some air circulation is beneficial for bottle gourd, strong winds can damage the vines and reduce fruit production. Providing a windbreak, such as a trellis or a fence, can help protect the crop.

By planting bottle gourd in suitable weather conditions, you can promote healthy growth, maximize fruit production, and enjoy a bountiful harvest.

3.5.4. Weather and Climatic Conditions Favorable for High Productivity

Bottle gourd is a warm-season crop that thrives in India's temperate climate. To achieve high productivity, it's essential to provide the crop with favorable weather and climate conditions.

- **Temperature:** Bottle gourd prefers a warm temperature, typically above 20°C (68°F). The ideal temperature range for growth and fruiting is between 25°C (77°F) and 32°C (90°F). Temperatures above 35°C (95°F) can lead to scorching, while temperatures below 15°C (59°F) can slow down growth.
- **Moisture:** Adequate moisture is crucial for bottle gourd growth and fruiting. The crop requires consistent soil moisture, especially during the germination and fruiting stages. However, excessive rainfall or waterlogging can lead to root rot and other diseases. An annual rainfall of 600-800 mm is considered ideal.

- **Sunlight**: Bottle gourd requires full sun to produce well, typically requiring at least 6 hours of direct sunlight per day.
- **Humidity**: Bottle gourd prefers a relatively high humidity, typically above 50%. However, high humidity can also lead to fungal diseases.
- **Wind**: While some air circulation is beneficial for bottle gourd, strong winds can damage the vines and reduce fruiting. Providing a windbreak, such as a trellis or a fence, can help protect the crop.
- **Region-wise** Climate Conditions:
    - **North India**: The plains of north India, with their warm and humid climate, are suitable for bottle gourd cultivation.
    - **South India**: The southern states, with their tropical climate, are also suitable for bottle gourd cultivation.

3.5.5. Effect of Extreme Weather Conditions

Bottle gourd is a warm-season crop that is sensitive to extreme weather conditions. Such conditions can significantly impact the crop's growth, fruiting, and overall productivity.

- **High Temperature**: High temperatures above 35°C (95°F) can cause scorching, sunburn, and heat stress, leading to reduced fruiting and lower yields. Prolonged exposure to high temperatures can also lead to the development of fungal diseases.
- **Low Temperature:** Low temperatures below 15°C (59°F) can slow down growth, reduce fruiting, and increase the risk of fungal diseases. Bottle gourd is highly susceptible to frost damage, which can cause significant yield losses.
- **Excessive Rainfall**: Excessive rainfall can cause waterlogging, leading to root rot and other diseases. It can also lead to nutrient leaching, reducing soil fertility and affecting plant growth.
- **Drought**: Drought can cause moisture stress, leading to reduced growth, fruiting, and yields. It can also increase the crop's susceptibility to pests and diseases.
- **Strong Winds**: Strong winds can damage the vines, reducing fruiting and yields. They can also cause soil erosion, reducing soil fertility and affecting plant growth.

To mitigate the effects of extreme weather conditions, farmers can use techniques such as mulching, irrigation, windbreaks, row covers, and disease management. These techniques can help reduce the impact of extreme weather conditions and promote healthy growth and productivity.

- **Effect of Hailstorms on Bottle Gourd:** Hailstorms can have a devastating impact on bottle gourd crops, causing significant damage and reducing yields.
  - **Physical Damage:** Hailstones can cause physical damage to the plants, including:
- **Bruising and crushing**: Hailstones can bruise and crush the leaves, stems, and fruits, leading to reduced growth and productivity.
  - **Defoliation**: Severe hailstorms can cause defoliation, leaving the plants vulnerable to disease and pest attacks.
- **Reduced Yields**: Hailstorms can reduce bottle gourd yields by:
  - **Damaging flowers and fruits**: Hailstones can damage the flowers and fruits, reducing the number of viable fruits and affecting the overall yield.
  - **Reducing plant growth**: Hailstorms can reduce plant growth, leading to smaller fruits and lower yields.
- **Increased Disease Susceptibility**: Hailstorms can increase the susceptibility of bottle gourd plants to diseases by:
  - **Causing wounds**: Hailstones can cause wounds on the plants, providing entry points for disease-causing pathogens.
  - **Reducing plant defenses**: Hailstorms can reduce the plant's natural defenses, making it more vulnerable to disease attacks.
- **Economic Losses**: Hailstorms can result in significant economic losses for bottle gourd farmers, including:
  - **Reduced income**: Hailstorms can reduce the income of farmers by decreasing the yield and quality of the crop.
  - **Increased costs**: Farmers may need to incur additional costs to repair or replace damaged plants, irrigation systems, and other infrastructure.

3.6. Bitter Gourd (*Momordica charantia*)

Bitter gourd also known as karela, is a popular vegetable in India, widely cultivated and consumed across the country (Fig. 3.6). Here's a detailed overview of its cultivation, production, and importance in India:

Fig. 3.6: Bitter gourd/karela crop

- **Area Under Cultivation**: Bitter gourd is grown in almost all states of India, with major production areas including Maharashtra, Gujarat, Uttar Pradesh, Bihar, and West Bengal. According to the Indian Ministry of Agriculture and Farmers Welfare, the area under bitter gourd cultivation in India has been steadily increasing over the years, with a total area of around 55,000 hectares under cultivation during 2020-21.
- **Production**: India is one of the largest producers of bitter gourd in the world. The country's production of bitter gourd has been increasing consistently, with a total production of around 1.8 million tonnes during 2020-21. Maharashtra, Gujarat, and Uttar Pradesh are the top three bitter gourd-producing states in India.
- **Importance**: Bitter gourd is an important crop in India, not only due to its nutritional value but also due to its cultural and economic significance. Here are a few reasons why bitter gourd is important in India:
  - Bitter gourd is a rich source of vitamins, minerals, and antioxidants, making it a nutritious addition to the Indian diet. It is particularly rich in vitamin C, vitamin B, and beta-carotene.
  - Bitter gourd has been used in traditional Indian medicine for centuries, particularly in Ayurveda. It is believed to have various health benefits, including reducing blood sugar levels, improving digestion, and boosting immunity.
  - Bitter gourd is an important crop for many small and marginal farmers in India, providing them with a source of income and livelihood. The crop is relatively easy to grow and requires minimal inputs, making it an attractive option for resource-poor farmers.
  - In addition to its nutritional and economic importance, bitter gourd also holds cultural significance in India. It is a popular ingredient in many traditional Indian recipes, particularly in North Indian and South Indian cuisine.

Overall, bitter gourd is an important crop in India, providing nutritional, economic, and cultural benefits to millions of people across the country. Its cultivation and consumption are likely to continue growing in the coming years, driven by increasing demand for nutritious and healthy food options.

3.6.1. Agroclimatic Conditions Suitable for Cultivation of Bitter Gourd

Bitter gourd is a warm-season crop that thrives in India's diverse agroclimatic conditions. Here's a detailed overview of the suitable conditions for growing bitter gourd in India:

- **Temperature**: Bitter gourd requires a warm temperature to grow, with optimal temperatures ranging from 24°C to 30°C. It can tolerate temperatures up to 35°C, but temperatures below 15°C can lead to poor growth and fruiting.
- **Soil:** Bitter gourd can be grown on a wide range of soils, but it prefers well-drained, fertile soils with a pH range of 6.0 to 6.7. It can tolerate some level of salinity, but extreme alkalinity or acidity can affect its growth.
- **Moisture**: Bitter gourd requires adequate moisture, especially during the fruiting stage. It thrives in areas with an annual rainfall of 600-800 mm, with supplemental irrigation during dry spells.
- **Sunlight**: Bitter gourd is a sun-loving crop that requires full sunlight to grow. It should be planted in areas that receive direct sunlight for at least 6 hours a day.
- **Region-wise suitability**: Bitter gourd can be grown in various agroclimatic regions of India, including:
  - The Indo-Gangetic Plains, which cover parts of Punjab, Haryana, Uttar Pradesh, Bihar, and West Bengal, are suitable for bitter gourd cultivation due to their warm and humid climate.
  - The Deccan Plateau, which covers parts of Maharashtra, Karnataka, and Andhra Pradesh, is also suitable for bitter gourd cultivation due to its warm and dry climate.
  - The Eastern Ghats, which cover parts of Odisha and Andhra Pradesh, are suitable for bitter gourd cultivation due to their warm and humid climate.
  - The Western Ghats, which cover parts of Maharashtra, Karnataka, and Kerala, are suitable for bitter gourd cultivation due to their warm and humid climate.

Overall, bitter gourd is a versatile crop that can be grown in various agroclimatic conditions across India.

3.6.2. Optimum Weather Conditions Suitable for Planting Bitter Gourd

The optimum weather conditions favorable for planting bitter gourd in India are:

Bitter gourd is typically planted in late spring to early summer, when the temperature is warm and the risk of frost has passed. The ideal time for planting bitter gourd in India is from February to April, or from June to November.

- **Temperature**: The ideal temperature for planting bitter gourd is between 25°C to 30°C. Temperatures above 35°C can lead to poor germination and seedling growth, while temperatures below 15°C can slow down growth.
- **Humidity**: Bitter gourd requires a relatively high humidity, typically above 60%, to thrive. However, excessive humidity can lead to fungal diseases.
- **Rainfall:**
- Bitter gourd requires adequate moisture, especially during the fruiting stage. It thrives in areas with an annual rainfall of 600-800 mm, with supplemental irrigation during dry spells.
- **Soil Temperature**: : The ideal soil temperature for planting bitter gourd is between 20°C to 25°C. Soil temperatures above 30°C can lead to poor germination, while soil temperatures below 15°C can slow down growth.
- **Sunlight:** Bitter gourd requires full sunlight to grow, with at least 6 hours of direct sunlight per day.
- **Wind**: Bitter gourd is sensitive to strong winds, which can damage the plants and reduce fruiting. It prefers a relatively calm environment, with wind speeds below 15 km/h.

3.6.3. Weather Conditions Favorable for High Productivity of Bitter Gourd

The weather conditions favorable for high productivity of bitter gourd in India are:

- **Temperature**: A warm temperature range of 25°C to 30°C is ideal for bitter gourd growth and productivity. Temperatures above 35°C can lead to flower drop and reduced fruiting, while temperatures below 15°C can slow down growth.

- **Relative Humidity**: A relative humidity of 60-80% is favorable for bitter gourd growth and productivity. High humidity can lead to fungal diseases, while low humidity can cause moisture stress.
- **Rainfall**: A well-distributed rainfall of 600-800 mm per annum is ideal for bitter gourd growth and productivity. Excessive rainfall can lead to waterlogged soil and reduced fruiting, while drought can cause moisture stress.
- **Sunshine:** Bitter gourd requires full sunlight to grow, with at least 6 hours of direct sunlight per day. Cloudy weather can reduce fruiting and productivity.
- **Wind:** A gentle breeze with wind speeds below 15 km/h is favorable for bitter gourd growth and productivity. Strong winds can damage the plants and reduce fruiting.
- **Soil Temperature**: A soil temperature range of 20°C to 25°C is ideal for bitter gourd growth and productivity. Soil temperatures above 30°C can lead to poor germination, while soil temperatures below 15°C can slow down growth.

3.6.4. Weather Conditions During Critical Stages

The following weather conditions are favorable during critical stages of bitter gourd growth:

- **During flowering**: Warm temperature (25-30°C), high relative humidity (60-80%), and gentle breeze.
- **During fruiting**: Warm temperature (25-30°C), moderate relative humidity (50-60%), and adequate moisture.
- **During seed formation**: Warm temperature (25-30°C), low relative humidity (40-50%), and dry weather.

3.6.5: Effect of Extreme Weather Events

Extreme weather events can significantly impact bitter gourd cultivation, affecting its growth, productivity, and quality. Drought, for instance, can lead to reduced plant growth and fruiting, increased susceptibility to pests and diseases, decreased fruit size and quality, and increased risk of flower drop and fruit abortion.

- Excessive rainfall or flooding can also be detrimental to bitter gourd crops. Waterlogged soil can lead to root rot and plant death, while reduced plant growth and fruiting can occur due to the excess moisture. Furthermore, the risk of fungal diseases increases, and fruit quality decreases due to the excessive moisture.

- Heatwaves can cause increased transpiration and water loss, leading to drought stress. This can result in reduced plant growth and fruiting, increased risk of sunscald and heat damage to fruits, and decreased fruit quality due to high temperatures.
- On the other hand, cold waves or frost can damage flowers and fruits, leading to reduced yields. Reduced plant growth and development can also occur, along with an increased risk of fungal diseases and decreased fruit quality due to cold damage.
- Strong winds can cause physical damage to plants and fruits, leading to reduced plant growth and fruiting. The risk of pest and disease outbreaks also increases, and fruit quality decreases due to wind damage. Hailstorms can cause similar physical damage to plants and fruits, leading to reduced yields and decreased fruit quality.

To mitigate the effects of extreme weather events on bitter gourd, farmers can adopt strategies such as using irrigation and mulching to conserve soil moisture, providing support to plants to prevent damage from strong winds, and implementing integrated pest management practices to reduce the risk of pest and disease outbreaks. Planting bitter gourd varieties that are tolerant to extreme weather conditions can also be an effective strategy.

3.7. Little Gourd (*Coccinia grandis*)

Little gourd, also known as Ivy gourd or baby watermelon, is a significant vegetable crop in India, particularly in the central Gujarat region (Fig. 3.7). Its importance lies in its nutritional value, medicinal properties, and export potential.

Fig. 3.7: Little gourd crop

- In terms of area and production, little gourd is widely cultivated in India, with the top-producing states being Gujarat, Maharashtra, and Uttar Pradesh. According to the National Horticulture Database (2019-20), India produced 191.77 million metric tons of vegetables, with little gourd being one of the prominent crops [1].
- The economic analysis of little gourd production in central Gujarat revealed that the average yield per hectare was around 327 quintals, with a gross income of Rs. 440,982 per hectare. The net profit per hectare was found to be Rs. 199,066, indicating the crop's profitability.
- Little gourd is not only a profitable crop but also possesses varied medicinal properties, making it a valuable resource for the pharmaceutical industry. Its export potential is also significant, with India being one of the major suppliers of little gourd to countries like the United States, the United Kingdom, and the Middle East.

To increase the area under little gourd cultivation and enhance its production, the government and farmers can focus on:

- **Improving irrigation facilities**: To reduce the crop's dependence on rainfall and increase its yield.
- **Adopting high-yielding varieties**: To enhance the crop's productivity and quality.
- **Providing training and support**: To farmers on best practices, pest management, and market trends.
- **Developing export-oriented infrastructure**: To facilitate the export of little gourd and increase its global market share.

3.7.1: Agroclimatic Conditions Favorable for Growing Little Gourd in India

Little gourd, also known as Ivy gourd or baby watermelon, is a warm-season crop that thrives in India's diverse agroclimatic conditions. The ideal conditions for growing little gourd in India is:

- Little gourd requires a warm temperature to grow, with optimal temperatures ranging from 25°C to 35°C. It can tolerate temperatures up to 40°C, but temperatures below 15°C can slow down growth.
- The crop requires adequate moisture, especially during the fruiting stage. A well-distributed rainfall of 600-800 mm per annum is ideal for little gourd growth and productivity. Excessive rainfall can lead to waterlogged soil and reduced fruiting.

- Little gourd prefers full sunlight to grow, with at least 6 hours of direct sunlight per day. Cloudy weather can reduce fruiting and productivity.
- The crop can be grown on a wide range of soils, but it prefers well-drained, fertile soils with a pH range of 6.0 to 7.0. It can tolerate some level of salinity, but extreme alkalinity or acidity can affect its growth.

In terms of region-wise suitability, little gourd can be grown in various agroclimatic regions of India, including the Indo-Gangetic Plains, the Deccan Plateau, and the Eastern and Western Ghats.

3.7.2. Weather Conditions Suitable for Planting

The ideal time for planting little gourd in India varies depending on the region.

- In general, it is planted in late spring to early summer, when the temperature is warm and the risk of frost has passed. In regions with a long growing season, it can be planted in the fall or early winter.
- Overall, little gourd is a versatile crop that can be grown in a variety of agroclimatic conditions across India. The optimum weather conditions required for high productivity of little gourd are warm temperatures, adequate moisture, and full sunlight. The ideal temperature range for little gourd growth and productivity is between 25°C to 35°C, with optimal temperatures for fruiting and flowering between 28°C to 32°C.
- Adequate moisture is also essential for high productivity, with a well-distributed rainfall of 600-800 mm per annum being ideal. The crop requires consistent moisture, especially during the fruiting stage, and excessive rainfall can lead to waterlogged soil and reduced fruiting.
- Full sunlight is also necessary for high productivity, with at least 6 hours of direct sunlight per day. Cloudy weather can reduce fruiting and productivity, and shaded conditions can lead to weak and spindly growth.
- In addition to these conditions, little gourd also requires a relatively high humidity, with an optimal range of 60-80%. Low humidity can lead to moisture stress, while high humidity can increase the risk of fungal diseases.
- Wind protection is also essential, as strong winds can damage the plants and reduce fruiting. A gentle breeze with wind speeds below 15 km/h is ideal for little gourd growth and productivity.

Overall, the optimum weather conditions for high productivity of little gourd require a delicate balance of warm temperatures, adequate moisture, full sunlight, and relatively high humidity.

3.7.3. Effect of Extreme Weather Events

Extreme weather events can significantly impact little gourd cultivation, affecting its growth, productivity, and quality. Here's a detailed overview of the effects of extreme weather events on little gourd:

- Drought can lead to reduced plant growth, lower fruit yields, and decreased fruit quality. Little gourd requires consistent moisture, especially during the fruiting stage, and drought can cause moisture stress, leading to reduced fruiting.
- Excessive rainfall or flooding can cause waterlogged soil, leading to root rot and plant death. Little gourd is sensitive to excess moisture, and prolonged waterlogging can reduce fruiting and productivity.
- Heatwaves can cause sunscald, heat damage, and reduced fruiting. Little gourd is sensitive to high temperatures, and prolonged exposure to temperatures above 35°C can lead to reduced fruit yields and lower fruit quality.
- Cold waves or frost can damage or kill little gourd plants, leading to reduced fruit yields and lower fruit quality. Little gourd is sensitive to low temperatures, and frost can cause significant damage to the crop.
- Strong winds can damage little gourd plants, reducing fruiting and productivity. Wind damage can also increase the risk of disease outbreaks and pest infestations.
- Hailstorms can cause physical damage to little gourd plants and fruits, leading to reduced fruit yields and lower fruit quality. Hail damage can also increase the risk of disease outbreaks and pest infestations.

To mitigate the effects of extreme weather events on little gourd, farmers can adopt strategies such as:

- Using irrigation and mulching to conserve soil moisture and reduce drought stress.
- Implementing drainage systems to prevent waterlogging and root rot.
- Providing shade and using cooling systems to reduce heat stress.
- Using row covers or other protective measures to prevent frost and cold damage.
- Providing support to plants to prevent wind damage.
- Using hail nets or other protective measures to prevent hail damage.

3.8. Pumpkins (*Cucurbita pepo*)

Pumpkins are a significant crop in India, widely cultivated and consumed across the country (Fig. 3.8). India is one of the largest producers of pumpkins in the world, and the crop plays an essential role in the country's agricultural sector.

Fig. 3.8: Pumpkin crop

- The total area under pumpkin cultivation in India was approximately 52,000 hectares in 2020-21, according to the Indian Ministry of Agriculture and Farmers Welfare. The major pumpkin-producing states in India are Uttar Pradesh, Maharashtra, Bihar, West Bengal, and Madhya Pradesh. These states account for a significant share of the country's total pumpkin production.
- In terms of production, India produced around 4.5 million metric tons of pumpkins in 2020-21. Pumpkins are typically grown in the kharif season (June to October) in India, and the crop takes around 100 to 120 days to mature after sowing. The ideal climate for pumpkin cultivation is warm and humid, with adequate rainfall and sunshine.
- Pumpkins are a popular vegetable in Indian cuisine, used in a variety of dishes such as curries, soups, and sweets. They are also rich in vitamins, minerals, and antioxidants, making them a nutritious addition to the Indian diet. In addition to their culinary and nutritional value, pumpkins also have significant cultural and economic importance in India.
- Pumpkin cultivation provides livelihood support to many farmers and rural households in India. The crop is often grown as a cash crop, providing a source of income for farmers. Additionally, pumpkins are also used in Indian festivals and traditions, such as the festival of lights, Diwali, where pumpkin lanterns are used as decorations.
- In recent years, there has been a growing trend towards organic and sustainable farming practices in India, and pumpkin cultivation is no

exception. Many farmers are adopting organic farming methods, which involve using natural pesticides and fertilizers, and avoiding genetically modified seeds. This approach not only helps to promote environmental sustainability but also contributes to improved crop quality and human health.

Overall, pumpkins are an important crop in India, providing a range of benefits, from culinary and nutritional value to economic and cultural significance. As the demand for pumpkins continues to grow, both domestically and internationally, India is well-positioned to meet this demand and capitalize on the opportunities presented by this versatile and nutritious crop.

3.8.1. Agroclimatic Conditions Favorable for Growing Pumpkins

Pumpkins are a warm-season crop that thrives in India's diverse agroclimatic conditions. To grow pumpkin successfully, specific conditions are required.

- Temperature is a crucial factor, with optimal temperatures ranging from 25°C to 30°C. Temperatures above 35°C can lead to sunscald and reduced fruit quality, while temperatures below 15°C can slow down growth and increase the risk of fungal diseases.
- Adequate moisture is also essential, especially during the fruiting stage. The ideal annual rainfall for pumpkin cultivation is around 600-800 mm. Irrigation is necessary in areas with low rainfall, with a water requirement of around 400-600 mm per crop season.
- Pumpkins prefer well-draining, fertile soils with a pH range of 6.0-6.8. The ideal soil texture is a mix of sand, silt, and clay, with good organic matter content. Full sun is also necessary, with a minimum of 6 hours of direct sunlight per day.
- Wind protection is vital, especially when the plants are young. Windbreaks, such as rows of trees or shrubs, can help reduce wind damage and promote healthy growth.
- A frost-free period of at least 100-120 days is necessary for pumpkins to mature. Areas with frequent frosts or cold snaps are not suitable for pumpkin cultivation.
- In terms of elevation, pumpkins can be grown up to 1,500 meters above sea level. However, high elevations can lead to cooler temperatures, affecting fruit quality and yield.
- One of the notable advantages of growing pumpkins is their long shelf life. When stored properly, pumpkins can last for several months, making them a valuable crop for farmers and consumers alike. This extended

shelf life allows for greater flexibility in marketing and distribution, enabling farmers to sell their produce over an extended period.

- In India, the ideal regions for growing pumpkins are the plains and valleys of the Indo-Gangetic belt, the Deccan Plateau, and the coastal regions of western and eastern India. States such as Uttar Pradesh, Maharashtra, Bihar, West Bengal, and Madhya Pradesh are well-suited for pumpkin cultivation.

By understanding the agroclimatic conditions required for growing pumpkins, farmers and growers can optimize their cultivation practices and improve crop yields and quality.

3.8.2. Optimum Weather Conditions for Sowing Pumpkins

For optimal growth and development, pumpkins require specific weather conditions during the sowing stage. Here are the ideal weather conditions for sowing pumpkins:

- **Temperature**: The ideal temperature for sowing pumpkins is between 20°C to 30°C. Temperatures above 35°C can lead to poor germination, while temperatures below 15°C can slow down germination.
- **Soil Temperature**: The ideal soil temperature for sowing pumpkins is between 18°C to 25°C.
- **Moisture:** The soil should be moist but not waterlogged. A consistent supply of moisture is essential for optimal germination.
- **Relative Humidity**: A relative humidity of 60-80% is ideal for sowing pumpkins.
- **Wind**: Gentle winds are beneficial for sowing pumpkins, as they help to distribute pollen and prevent fungal diseases.

In India, the ideal time for sowing pumpkins varies depending on the region:

- In the northern states, such as Uttar Pradesh, Punjab, and Haryana, the ideal time for sowing pumpkins is from mid-February to mid-March.
- In the southern states, such as Tamil Nadu, Karnataka, and Andhra Pradesh, the ideal time for sowing pumpkins is from mid-January to mid-February.
- In the eastern states, such as West Bengal, Odisha, and Bihar, the ideal time for sowing pumpkins is from mid-February to mid-March.

By sowing pumpkins during the optimal weather conditions, farmers can ensure good germination, healthy seedling growth, and a higher yield.

3.8.3. Agroclimatic Conditions Favorable for High Productivity of Pumpkins

Pumpkins are a warm-season crop that thrives in India's diverse agroclimatic conditions. To achieve high productivity, pumpkins require a specific set of agroclimatic conditions.

- Temperature plays a crucial role in pumpkin cultivation. The ideal temperature for pumpkin growth ranges from 25°C to 30°C. Temperatures above 35°C can lead to sunscald and reduced fruit quality, while temperatures below 15°C can slow down growth and increase the risk of fungal diseases.
- Adequate moisture is also essential for high pumpkin productivity. The ideal annual rainfall for pumpkin cultivation is around 600-800 mm. Irrigation is necessary in areas with low rainfall, with a water requirement of around 400-600 mm per crop season.
- Pumpkins prefer well-draining, fertile soils with a pH range of 6.0-6.8. The ideal soil texture is a mix of sand, silt, and clay, with good organic matter content.
- Full sun is necessary for optimal pumpkin growth, with a minimum of 6 hours of direct sunlight per day. Wind protection is also vital, especially when the plants are young. Windbreaks, such as rows of trees or shrubs, can help reduce wind damage and promote healthy growth.
- A frost-free period of at least 100-120 days is necessary for pumpkins to mature. Areas with frequent frosts or cold snaps are not suitable for pumpkin cultivation.
- In terms of elevation, pumpkins can be grown up to 1,500 meters above sea level. However, high elevations can lead to cooler temperatures, affecting fruit quality and yield.
- The ideal agroclimatic conditions for high pumpkin productivity can be found in the plains and valleys of the Indo-Gangetic belt, the Deccan Plateau, and the coastal regions of western and eastern India. States such as Uttar Pradesh, Maharashtra, Bihar, West Bengal, and Madhya Pradesh are well-suited for pumpkin cultivation.

By understanding the agroclimatic conditions required for high pumpkin productivity, farmers and growers can optimize their cultivation practices and improve crop yields and quality.

3.8.4. Effect of Extreme Weather Events

Extreme weather events can have a significant impact on pumpkin crops, affecting their growth, development, and productivity. Here are some of the ways extreme weather events can affect pumpkins:

- **Drought**: Drought can lead to water stress, causing pumpkins to become misshapen, smaller, and less flavorful. Prolonged drought can also increase the risk of fungal diseases.
- **Heavy Rainfall:** Excessive rainfall can lead to waterlogged soils, causing roots to rot and reducing plant growth. Heavy rainfall can also increase the risk of fungal diseases and nutrient deficiencies.
- **Heatwaves**: High temperatures above 35°C can cause sunscald, leading to white or yellowish patches on the fruit. Heat stress can also reduce fruit quality and increase the risk of fungal diseases.
- **Floods:** Flooding can cause roots to rot, leading to plant death. Floodwaters can also carry diseases and pests that can infect pumpkin plants.
- **Hailstorms:** Hail can damage pumpkin fruits, causing them to become bruised or broken. Hail can also damage leaves and stems, reducing plant growth.
- **Strong Winds**: Strong winds can damage pumpkin plants, causing them to become lodged or broken. Winds can also increase the risk of fungal diseases and reduce fruit quality.
- **Temperature Fluctuations**: Temperature fluctuations can affect pumpkin growth and development. For example, a sudden drop in temperature can cause fruit to become misshapen or develop off-flavors.

To mitigate the effects of extreme weather events on pumpkins, farmers can use various strategies such as:

- Irrigation management to reduce water stress during droughts
- Drainage management to reduce waterlogging during heavy rainfall
- Shade management to reduce heat stress during heatwaves
- Flood protection measures such as levees or floodwalls
- Windbreaks such as rows of trees or shrubs to reduce wind damage
- Crop insurance to manage risk and uncertainty

By understanding the effects of extreme weather events on pumpkins, farmers and growers can take proactive steps to mitigate these impacts and ensure a healthy and productive crop Pumpkins are typically grown as a spreading crop, and they can be cultivated without support.

3.8.5. Managing the Spreading of Pumpkin

The vines of the pumpkin plant spread out on the ground, covering a large area. This type of growth habit is known as a "spreading" or "trailing" habit.

Growing pumpkins without support allows the vines to spread out freely, which can help to:

- Suppress weeds by shading the soil
- Retain moisture in the soil
- Regulate soil temperature

However, growing pumpkins without support can also have some drawbacks, such as:

- Reduced air circulation, which can increase the risk of fungal diseases
- Increased soil moisture, which can lead to rot and other problems
- Difficulty in accessing the fruit, as it may be hidden under the sprawling vines

To overcome these challenges, farmers and gardeners can use various techniques, such as:

- Mulching to retain moisture and suppress weeds
- Pruning to improve air circulation and promote healthy growth
- Using a trellis or other support system to lift the vines off the ground and improve access to the fruit.

3.9. Ash Gourd (*Benincasa hispida*)

Ash gourd, also known as white gourd or winter melon, is a significant vegetable crop in India (Fig. 3.9). Here's an overview of its area, production, and importance:

- Ash gourd is cultivated in many parts of India, with major producing states including Maharashtra, Gujarat, Uttar Pradesh, and Tamil Nadu. According to the Indian Horticultural Database, the total area under ash gourd cultivation in India is around 130,000 hectares.
- In terms of production, India is one of the largest producers of ash gourd globally. As per the data from the National Horticultural Board, the annual production of ash gourd in India is approximately 3.5 million metric tons.

Fig. 3.9: Ash gourd crop

- Ash gourd holds significant importance in Indian cuisine, culture, and traditional medicine. It is a versatile vegetable used in various dishes, such as soups, curries, and sweets. The gourd is also used in Ayurvedic medicine for its perceived health benefits, including reducing inflammation and improving digestion. Additionally, ash gourd is considered a sacred vegetable in some Hindu traditions and is used in various rituals and ceremonies. Notably, ash gourd is also used to make Agra Petha, a popular Indian sweet dish that originated in Agra. India

3.9.1. Agroclimatic Conditions Required for Growing Ash Gourd

Ash gourd is a warm-season crop that thrives in India's diverse agroclimatic conditions. Here are the suitable conditions for growing ash gourd in India:

- Ash gourd requires a warm and humid climate with temperatures ranging from 25°C to 35°C. It can tolerate a maximum temperature of 40°C and a minimum temperature of 10°C.
- The ideal relative humidity for ash gourd cultivation is between 60% to 80%. It requires adequate moisture, especially during the fruiting stage.
- Ash gourd can be grown in a wide range of soils, but it prefers well-draining, rich loamy soils with a pH range of 6.0 to 7.0.
- Ash gourd requires adequate rainfall, with an annual rainfall of at least 600 mm. However, it is susceptible to excessive rainfall, which can lead to fungal diseases.

3.9.2: Favorable Periods for Sowing

The ideal time for sowing ash gourd seeds in India varies depending on the region:

- In northern India, the best time for sowing is from April to June.
- In southern India, the ideal time for sowing is from February to April.
- In eastern and western India, the best time for sowing is from March to May.

By considering these agroclimatic conditions, farmers can optimize ash gourd cultivation in India.

3.9.3. Weather Conditions Favorable for High Productivity

For high productivity of ash gourd in India, the following weather conditions are favorable:

- Ash gourd requires a warm temperature range of 25°C to 32°C. Temperatures above 35°C can lead to sun scorching and reduced yields.
- A well-distributed annual rainfall of 800-1000 mm is favorable for high productivity. Consistent moisture, especially during the fruiting stage, is essential.
- A relative humidity of 60-80% is ideal for ash gourd growth. High humidity helps in better fruit set and development.
- Ash gourd requires full sun to produce well, so areas receiving direct sunlight for at least 6 hours a day are suitable. Wind speed should be less than 15 km/h, as strong winds can damage plants and reduce yields.
- Consistent soil moisture, especially during critical stages of growth (flowering and fruiting), is essential for high productivity.

In India, regions with favorable weather conditions for high productivity of ash gourd include the plains of northern India, western states, and southern states. These regions typically experience warm temperatures, adequate rainfall, and sufficient sunshine.

3.9.4: Management of Spreading

Ash gourds have long shelf life Managing the spreading of ash gourd creepers is crucial to optimize space, promote healthy growth, and increase yields. Here's a step-by-step guide:

- Ash gourd creepers can spread quite aggressively, so providing a trellis or a support system is essential. Set up a sturdy trellis or a fence around the planting area, about 1-2 meters high.
- Once the creepers start spreading, gently twine them around the trellis or support system. This will help to keep the creepers organized, promote vertical growth, and make it easier to manage the plants.
- As the creepers grow, continue to twine them around the trellis, making sure to leave enough space between each creeper for proper air circulation and sunlight penetration.
- To further manage the spreading of the creepers, you can also use a technique called "pruning". Prune the tips of the creepers to encourage branching and prevent them from spreading too far.

By providing support, twining the creepers, and pruning, you can effectively manage the spreading of ash gourd creepers, optimize space, and promote healthy growth.

3.9.5. Effect of Extreme Weather Events

Extreme weather events can have a significant impact on ash gourd cultivation, affecting its growth, productivity, and quality. Here's a detailed overview of the effects of extreme weather events on ash gourd:

- **Drought** is one of the most significant extreme weather events affecting ash gourd. Ash gourd requires consistent moisture, especially during the fruiting stage. Prolonged drought can lead to reduced plant growth, lower fruit set, and decreased yields. Drought stress can also make the plants more susceptible to pests and diseases.
- On the other hand, **excessive rainfall** can also be detrimental to ash gourd. Waterlogged soil can lead to root rot, reducing plant growth and productivity. Excessive rainfall can also increase the incidence of fungal diseases, such as powdery mildew and downy mildew.
- **High temperatures**, typically above 35°C, can cause sun scorching and reduce yields. Ash gourd is sensitive to heat stress, which can lead to reduced fruit set, lower fruit quality, and increased susceptibility to pests and diseases.
- **Flooding** is another extreme weather event that can significantly impact ash gourd. Flooded soil can lead to root rot, reducing plant growth and productivity. Flooding can also increase the incidence of fungal diseases and reduce the effectiveness of pesticides.
- **Hailstorms** can cause physical damage to ash gourd plants, reducing yields and affecting fruit quality. Hail damage can also create entry points for pests and diseases.
- **Windstorms** can also impact ash gourd, causing physical damage to plants and reducing yields. Strong winds can also disrupt pollination, reducing fruit set.

To mitigate the effects of extreme weather events on ash gourd, farmers can adopt various strategies, such as using drought-tolerant varieties, implementing irrigation management practices, and providing support to plants to prevent damage from wind and hail.

3.10. Tindas (*Benincasa fistulosa*)

Tinda, also known as baby pumpkin or Indian squash, is a popular vegetable crop in India (Fig. 3.10).

Fig. 3.10: Tinda crop

Here's an overview of its area, production, and importance:

- Tinda is cultivated in many parts of India, with major producing states including Punjab, Uttar Pradesh, Bihar, and Maharashtra. According to the Indian Horticultural Database, the total area under Tinda cultivation in India is around 120,000 hectares.
- In terms of production, India is one of the largest producers of Tinda globally. As per the data from the National Horticultural Board, the annual production of Tinda in India is approximately 2.5 million metric tons.
- Tinda holds significant importance in Indian cuisine and culture. It is a versatile vegetable used in various dishes, such as curries, soups, and stir-fries. Tinda is also used in traditional Indian medicine, particularly in Ayurveda, for its perceived health benefits, including reducing inflammation and improving digestion.
- In addition to its culinary and medicinal uses, Tinda is also an important crop for farmers in India. It is a low-input crop that requires minimal irrigation and fertilizers, making it an attractive option for small-scale farmers. Tinda is also a good source of income for farmers, particularly during the off-season when other vegetable prices are low.
- Furthermore, Tinda is a nutrient-rich vegetable, providing essential vitamins, minerals, and antioxidants. It is an excellent source of vitamin C, vitamin K, and potassium, making it a healthy addition to a balanced diet.

Overall, Tinda is an important crop in India, providing nutritional, economic, and cultural benefits to farmers, consumers, and the country as a whole.

3.10.1. Agroclimatic Conditions Required for Growing Tindas

Tinda requires a warm and humid climate with temperatures ranging from 20°C to 35°C.

- The ideal temperature for germination is around 25°C to 30°C, while fruiting requires a temperature range of 20°C to 25°C.
- In terms of rainfall, Tinda requires adequate moisture, especially during the fruiting stage. An annual rainfall of around 600 mm to 800 mm is considered ideal for Tinda cultivation. However, the crop is sensitive to excessive rainfall, which can lead to fungal diseases.
- Tinda can be grown in a variety of soils, but it prefers well-draining, sandy loam soils with a pH range of 6.0 to 7.0. The crop is sensitive to waterlogging, so soils with good drainage are essential.
- In India, Tinda is typically grown during the kharif season, which coincides with the monsoon season. Tinda requires a warm and humid climate with adequate rainfall and well-draining soils to grow well in India. Successful Tinda cultivation relies on providing the right climate conditions throughout the plant's growth cycle.

By understanding and optimizing factors such as temperature, soil quality, watering, humidity, and protection from pests and diseases, growers can ensure a healthy and productive Tinda crop. Whether you are a seasoned farmer or a backyard gardener, paying attention to these climate requirements will contribute to the success of your Tinda cultivation endeavors.

3.10.2. Optimum Weather Conditions for High Productivity

To successfully cultivate Tinda, it is crucial to understand and provide the optimal climate conditions for its growth. In this article, we will explore the climate requirements for successful Tinda cultivation.

- **Temperature:** Tinda thrives in warm and tropical climates, making it well-suited for regions with temperatures ranging between 20°C to 35°C (68°F to 95°F). While it can tolerate high temperatures, it is essential to avoid extreme heat or cold, as these conditions can adversely affect its growth.
- **Sunlight**: Adequate sunlight is crucial for Tinda plants, so selecting a location with full sunlight exposure is recommended.
- **Moisture:** Tinda plants require modest watering throughout their growth cycle. It is important to keep the soil evenly moist but not waterlogged. Excess water can lead to root rot and other diseases, while moisture stress can result in stunted growth and reduced yield. Tinda plants grow

better when relative humidity is between 60 to 80 per cent .Adequate humidity is particularly important during the flowering and fruit-setting stages to ensure successful pollination and fruit development.

- **Season:** Tinda is a warm-season crop, and it is typically grown during the summer and monsoon seasons. The ideal time for sowing Tinda seeds is after the peak winter period and the soil has warmed up sufficiently.

3.10.3. Effect of Extreme Weather Events on Tindas

Continuous heavy rains with flooding will cause severe damage to the crop. Drought and moisture also will result in very low yields or crop failure.

3.11. Parwal (*Trichosanthes dioica* Roxb.)

Pointed Gourd also known as Parwal is one of the important vegetable of India (Fig. 3.11).

Fig. 3.11: Parwal crop

It is mainly cultivated in the states of Uttar Pradesh, Bihar, Assam, West Bengal, Odisha, Madhya Pradesh, Maharashtra, Gujarat, Kerala and Tamil Nadu. The fruits of pointed gourd are rich in proteins and vitamin A. Additionally it possess medicinal properties that can help to lower blood sugar and serum triglycerides. In India, it is cultivated in an area about 20000 ha with production of 310000 metric tons and productivity of 14.89 t/ha during the year 2018-2019.

3.11.1. Agroclimatic Requirements

It is a day-neutral, perennial plant, thrives well in regions with moderately warm to hot humid climates and sensitive to frost and severe cold, the ideal temperature for this plant growth is between 30-35^{0}C. It cannot survive in waterlogged conditions. Sandy loam soil with a pH level of 6.0–6.5 is best suited for cultivation of Parwal.

- **Planting Time**: The appropriate time for the planting of Parwal in the plains is from mid- July to October while in the Diara area, the ideal time for planting is from September to October.
- **Propagation**: Parwal can be propagate through various methods, with seeds, root cuttings and vine being the primary methods.
- **Propagation through root cutting**: In this method, 1 to 2 inch part of the stem along with the roots, containing 5-6 nodes, is planted. This method results in rapid plants growth, and fruits start appearing earlier than usual.
- **Propagation through vine cutting**: In this method, planting is done by vine cuttings, and the length of the vines should be 120-150 cm and one year old. Cultivation of pointed gourd is carried out on a large scale in Uttar Pradesh and Bihar using these methods.

Pointed gourd is a tropical vegetable crop that is sensitive to extreme weather events. Here are some potential effects of extreme weather events on pointed gourd:

3.11.2. Extreme Weather Events and Their Effects

- **Drought**: Reduced yield and fruit quality due to water stress. Increased susceptibility to pests and diseases. Decreased vine growth and fruit set.
- **Floods**: Root rot and plant death due to waterlogged soil. Reduced oxygen availability, leading to plant stress. Increased risk of fungal diseases.
- **Heatwaves**: Sun scorching and fruit damage due to high temperatures. Reduced pollination and fruit set. Increased water loss, leading to drought stress.
- **Cyclones and Strong Winds**: Physical damage to plants and vines. Reduced fruit quality and yield due to mechanical stress. Increased risk of disease transmission.

3.12. Sponge Gourd (*Luffa cylindrica* syn. L. aegyptiaca)

Sponge gourd is another important vegetable, widely grown in India. It is very similar to ridge gourd. Brazil, India, Mexico and Ghana are the country where sponge gourd is grown generally (Fig, 3.12). Gilki and Turai is the most common name of sponge gourd in India.

Fig, 3.12: Sponge gourd crop

- Sponge gourd is very rich in vitamin A and C. Seed oil of sponge gourd is helpful to cure skin diseases. Juice of sponge gourd is very effective in curing diabetes.

Climatic Requirements

- Warm and humid climate is suitable for its cultivation.
- Required temperature for the cultivation should be in between 24-37^{0}C
- At the stage of flowering and fruiting, excessive rainfall can cause poor yield.
- Too low temperature can cause delay in germination and retarded early growth.

Soils

- pH of soil should be in range of 5.5-6.5.
- Although cultivation of sponge gourd can take place in all type of soils but soil should have good moisture holding capacity especially in summer.
- Soil should be rich in organic matter.
- Sandy loam soil is comparatively best for the cultivation of sponge gourd.

Region-wise Weather Conditions

- In North India, sponge gourd is grown in the summer season (April-June) when the temperature is between 25-35°C (77-95°F).

- In South India, it is grown in the summer season (February-April) when the temperature is between 25-30°C (77-86°F).
- In East India, it is grown in the summer season (March-May) when the temperature is between 25-30°C (77-82°F).

By providing ridge gourd plants with these optimum weather conditions, you can promote healthy growth, high productivity, and better fruit quality in India.

3.13. Conclusion

- Agroclimatology plays a vital role in understanding the growth and development of creeper-type vegetables in India. The country's diverse climate and geographical regions demand a thorough understanding of the complex interactions between weather, soil, and crop. By studying agroclimatology, farmers and agricultural scientists can develop strategies to mitigate the impacts of extreme weather events, optimize crop yields, and improve the overall sustainability of vegetable production.
- Furthermore, agroclimatology can help identify suitable regions and climates for specific types of creeper vegetables, enabling farmers to make informed decisions about crop selection and management. By integrating agroclimatology with other disciplines, such as plant breeding, soil science, and entomology, India can develop a more resilient and productive vegetable production system. This, in turn, can contribute to the country's food security, economic growth, and sustainable development.

4

Agroclimatology of Different Types of Beans Used as Vegetables

Beans play a vital role in India's agricultural sector and food security. They are an essential source of protein, fiber, and other essential nutrients, making them a crucial component of the Indian diet. Beans are also a significant source of income for many farmers in India, particularly small and marginal farmers. The country's bean production also contributes significantly to the national economy, with India being one of the largest producers of beans in the world. In addition to their nutritional and economic importance, beans also play a crucial role in maintaining soil fertility and promoting sustainable agriculture practices in India. Beans are legumes, which mean they have the ability to fix nitrogen in the soil, reducing the need for synthetic fertilizers and promoting soil health. India is a significant producer of various types of beans, which are widely used as vegetables in the country. These beans are not only a good source of protein, fiber, and other essential nutrients but also play a crucial role in the country's agricultural sector.

Overall, beans are a vital crop in India, and their production and consumption play a significant role in the country's food security, agricultural sector, and national economy.

4.1. Kidney Beans (*Phaseolus vulgaris*)

Kidney beans, also known as Rajma are one of the most widely cultivated and consumed beans in India (Fig. 4.1).

Fig. 4.1: Kidney beans crop

They are grown in various parts of the country, with major production areas including Punjab, Haryana, and Uttar Pradesh. According to the Indian Ministry of Agriculture and Farmers Welfare, the country produced over 1.4 million metric tons of kidney beans in 2020-21. Kidney beans are an essential ingredient in many popular Indian dishes, such as rajma masala, and are also used in salads, soups, and stews.

- The kidney bean-growing regions in India are characterized by a subtropical climate with warm summers and cool winters. The major kidney bean-producing states, including Punjab, Haryana, and Uttar Pradesh, have a continental climate with extreme temperatures.
- The temperature requirements for kidney bean cultivation are moderate, with optimal temperatures ranging from 18°C to 30°C. The crop is sensitive to extreme temperatures, and temperatures above 35°C can lead to flower drop and reduced pod formation.
- The kidney bean-growing regions receive an average annual rainfall of 600-800 mm, with the majority of the rainfall occurring during the monsoon season. The crop requires adequate moisture during the germination and flowering stages, and drought can significantly impact yields.

4.1.1. Weather Conditions Suitable for Sowing

The ideal weather conditions for sowing kidney beans are during the winter season, typically from October to November. At this time, the weather is characterized by mild temperatures, low humidity, and adequate moisture in the soil.

- The optimal temperature for sowing kidney beans ranges from 15°C to 20°C. Temperatures above 25°C can lead to poor germination and reduced seedling growth.
- Adequate moisture in the soil is also essential for sowing kidney beans. The soil should be moist but not waterlogged, with a minimum of 20-25 cm of rainfall or irrigation before sowing.
- Low humidity is also beneficial for sowing kidney beans, as high humidity can lead to fungal diseases. A relative humidity of 60-70% is ideal for sowing kidney beans.

In addition to these weather conditions, the timing of sowing is also critical. In India, the best time to sow kidney beans is during the rabi season, which typically starts in October and lasts until March. Sowing during this period allows the crop to take advantage of the cool winter weather and adequate moisture, resulting in better yields and quality.

4.1.2. Optimum Weather Conditions Required for High Productivity

In India, the optimum weather conditions for high productivity of kidney beans are characterized by a combination of temperature, moisture, and sunshine.

- Temperature is a critical factor, and kidney beans thrive in temperatures ranging from 15°C to 25°C. Optimal temperatures for germination, growth, and pod formation are 18°C to 22°C. Temperatures above 30°C can lead to flower drop, reduced pod formation, and lower yields.
- Adequate moisture is essential for kidney bean production, and the crop requires consistent and well-distributed rainfall or irrigation throughout the growing season. The ideal annual rainfall for kidney beans is between 600-800 mm, with a minimum of 20-25 cm of rainfall or irrigation during the critical growth stages.
- Sunshine is also crucial for kidney bean production, and the crop requires full sun to produce well. Bright sunshine during the flowering and pod-formation stages is particularly important for promoting healthy growth and maximizing yields.
- Soil moisture is also critical, and kidney beans prefer well-drained, fertile soils with a pH range of 6.0-7.0. The crop is sensitive to waterlogging, and excessive moisture can lead to root rot and reduced yields.

4.1.3. Effect of Extreme Weather Events

Extreme weather events can have a significant impact on kidney bean crops in India. The country's agricultural sector is highly vulnerable to climate-related shocks, and kidney beans are no exception.

- Drought is one of the most significant extreme weather events affecting kidney bean crops. Insufficient rainfall during the critical growth stages can lead to reduced yields, poor quality, and even complete crop failure. Kidney beans require consistent and well-distributed rainfall throughout the growing season, and drought can disrupt this process. In addition, drought can also lead to increased susceptibility to pests and diseases, further exacerbating the damage to the crop.
- Heatwaves are another extreme weather event that can significantly impact kidney bean crops. High temperatures above 35°C can lead to flower drop, reduced pod formation, and lower yields. Heat stress can also cause damage to the crop's reproductive structures, leading to reduced seed quality and viability. Furthermore, heatwaves can also increase the spread of pests and diseases, further damaging the crop.
- Flooding and waterlogging are also extreme weather events that can significantly impact kidney bean crops. Excessive rainfall during the

critical growth stages can lead to waterlogging, which can cause root rot and reduced yields. Kidney beans are sensitive to waterlogging, and excessive moisture can lead to a range of problems, including reduced seed germination, poor seedling growth, and increased susceptibility to pests and diseases.

- Hailstorms are another extreme weather event that can significantly impact kidney bean crops. Hail can cause physical damage to the crop, including bruising, breaking, and shattering of the pods. This can lead to reduced yields, poor quality, and even complete crop failure.
- In addition to these extreme weather events, climate change is also expected to have a significant impact on kidney bean crops in India. Rising temperatures, changing precipitation patterns, and increased frequency of extreme weather events are all expected to affect the crop's growth, yield, and quality.

To mitigate the impact of extreme weather events on kidney bean crops, farmers in India are adopting a range of strategies, including the use of drought-tolerant and heat-resistant varieties, conservation agriculture practices, and integrated pest management techniques. Additionally, governments and other stakeholders are also providing support to farmers through initiatives such as crop insurance programs, weather-based crop insurance, and climate-resilient agriculture programs.

4.2. French Beans (*Phaseolus vulgaris*)

French beans, also known as green beans, are another popular variety of beans grown in India (Fig. 4.2).

Fig. 4.2: French beans crop

They are widely cultivated in states like Maharashtra, Gujarat, and Karnataka. The country produced over 0.6 million metric tons of French beans in 2020-

21. French beans are a versatile ingredient and are used in various dishes, including stir-fries, salads, and curries.

- According to the Indian Ministry of Agriculture and Farmers Welfare, the country produced over 0.6 million metric tons of French beans in 2020-21, with a total area of over 0.3 million hectares under cultivation. The major French bean-producing states in India are Maharashtra, Gujarat, and Karnataka, which together account for over 60% of the country's total French bean production.
- French beans are an important crop in India, providing a valuable source of income for many farmers, particularly small and marginal farmers. The crop is also a significant contributor to the country's vegetable production, with French beans being one of the top five vegetable crops in terms of production.
- In addition to their economic importance, French beans are also a nutritious and versatile crop, rich in vitamins, minerals, and antioxidants. They are a popular ingredient in many Indian dishes, including stir-fries, curries, and salads.

The importance of French beans in India can be gauged from the following facts:

- French beans are a significant export commodity for India, with the country exporting over 10,000 metric tons of French beans in 2020-21.
- French beans are a valuable source of employment for rural communities, particularly during the harvesting and processing seasons.
- The crop is also an important part of India's food security, providing a nutritious and affordable source of protein and fiber for millions of people.

Overall, French beans are a vital crop in India, providing economic, nutritional, and employment benefits to millions of people.

4.2.1. Agroclimatic Characterstics of Regions Growing French Beans

The French bean-growing regions in India are characterized by a subtropical climate with warm summers and mild winters. The major French bean-producing states, including Maharashtra, Gujarat, Karnataka, and Tamil Nadu, have a diverse range of agroclimatic conditions.

- Temperature is an important factor in French bean cultivation, and the crop thrives in temperatures ranging from 15°C to 30°C. The ideal temperature for germination and seedling growth is between 20°C to 25°C, while temperatures above 30°C can lead to flower drop and reduced pod formation.

- Moisture is another critical factor, and French beans require adequate and well-distributed rainfall throughout the growing season. The ideal annual rainfall for French beans is between 600-800 mm, with a minimum of 20-25 cm of rainfall or irrigation during the critical growth stages.
- Soil is also an important factor, and French beans prefer well-drained, fertile soils with a pH range of 6.0-7.0. The crop can be grown on a variety of soils, including loam, clay, and sandy loam.
- In terms of sunshine, French beans require full sun to produce well, and bright sunshine during the flowering and pod-formation stages is particularly important for promoting healthy growth and maximizing yields.
- The French bean-growing regions in India experience a range of agroclimatic conditions, including tropical, subtropical, and temperate climates. The crop is typically grown during the rabi season (October-March) in the northern states, while in the southern states, it is grown during the kharif season (June-September).

Overall, the agroclimatic conditions in the French bean-growing regions of India are diverse, and the crop is adapted to a range of temperature, moisture, and soil conditions.

4.2.2. Optimum Weather Conditions for High Productivity

The optimum weather conditions required for high productivity of French beans include a combination of temperature, moisture, and sunshine.

- Temperature is a critical factor, and French beans thrive in temperatures ranging from 15°C to 30°C. The ideal temperature for germination and seedling growth is between 20°C to 25°C. Temperatures above 30°C can lead to flower drop, reduced pod formation, and lower yields.
- Adequate moisture is essential for French bean production, and the crop requires consistent and well-distributed rainfall throughout the growing season. The ideal annual rainfall for French beans is between 600-800 mm, with a minimum of 20-25 cm of rainfall or irrigation during the critical growth stages.
- French beans require full sun to produce well, and bright sunshine during the flowering and pod-formation stages is particularly important for promoting healthy growth and maximizing yields. A minimum of 6 hours of direct sunshine per day is recommended for optimal French bean production.

- Soil temperature is also an important factor, and French beans prefer a soil temperature of around 20°C to 25°C for optimal germination and growth.
- In terms of humidity, French beans prefer a relative humidity of 60-70% during the flowering and pod-formation stages. High humidity can lead to fungal diseases, while low humidity can cause moisture stress.
- Wind protection is also essential for French bean production, as strong winds can damage the crop and reduce yields.

Overall, the optimum weather conditions required for high productivity of French beans include moderate temperatures, adequate moisture, full sun, and protection from strong winds.

4.2.3. Effect of Extreme Weather Conditions on the Crop

Extreme weather conditions can have a significant impact on French beans crops, affecting their growth, yield, and quality.

- Drought is one of the most significant extreme weather conditions affecting French beans. Insufficient rainfall during the critical growth stages can lead to reduced yields, poor quality, and even complete crop failure. French beans require consistent and well-distributed rainfall throughout the growing season, and drought can disrupt this process.
- Heatwaves are another extreme weather condition that can significantly impact French beans crops. High temperatures above 35°C can lead to flower drop, reduced pod formation, and lower yields. Heat stress can also cause damage to the crop's reproductive structures, leading to reduced seed quality and viability.
- Flooding and waterlogging are also extreme weather conditions that can significantly impact French beans crops. Excessive rainfall during the critical growth stages can lead to waterlogging, which can cause root rot and reduced yields. French beans are sensitive to waterlogging, and excessive moisture can lead to a range of problems, including reduced seed germination, poor seedling growth, and increased susceptibility to pests and diseases.
- Hailstorms are another extreme weather condition that can significantly impact French beans crops. Hail can cause physical damage to the crop, including bruising, breaking, and shattering of the pods. This can lead to reduced yields, poor quality, and even complete crop failure.
- Strong winds are also an extreme weather condition that can significantly impact French beans crops. Strong winds can cause physical damage to

the crop, including lodging and breakage of the plants. This can lead to reduced yields, poor quality, and even complete crop failure.

- Climate change is also expected to have a significant impact on French beans crops in India. Rising temperatures, changing precipitation patterns, and increased frequency of extreme weather events are all expected to affect the crop's growth, yield, and quality.

To mitigate the impact of extreme weather conditions on French beans crops, farmers in India are adopting a range of strategies, including the use of drought-tolerant and heat-resistant varieties, conservation agriculture practices, and integrated pest management techniques. Additionally, governments and other stakeholders are also providing support to farmers through initiatives such as crop insurance programs, weather-based crop insurance, and climate-resilient agriculture programs.

4.3. Cluster Beans (*Cyamopsis tetragonoloba*)

Cluster beans, also known as guar beans, are a type of bean commonly grown in India (Fig. 4.3).

Fig. 4.3: Cluster bean crop

They are primarily cultivated in states like Rajasthan, Gujarat, and Maharashtra. The country produced over 0.4 million metric tons of cluster beans in 2020-21. Cluster beans are used in various dishes, including curries, stews, and salads. The crop is usually sown in July-August and harvested in October-November.

- According to the Ministry of Agriculture and Farmers Welfare, the area under cluster bean cultivation in India has been steadily increasing over the years. In 2020-21, the area under cluster bean cultivation was approximately 3.5 million hectares.

- In terms of production, India is the largest producer of cluster beans, accounting for over 80% of global production. The country produced around 6.5 million tonnes of cluster beans in 2020-21.
- Cluster beans are an important crop in India due to their multiple uses. The beans are a rich source of protein, fiber, and minerals, making them a nutritious food source. They are also used as animal feed, and the guar gum extracted from the beans is used in various industries, including food, pharmaceuticals, and cosmetics. Additionally, cluster beans are a valuable cash crop for many farmers in India, providing them with a source of income.

4.3.1. Agroclimatic Characstics of Cluster Beans Growing Regions

Cluster beans are typically grown in regions with specific agroclimatic characteristics. These regions are primarily located in the northwestern and western parts of India.

- The ideal climate for cluster bean cultivation is characterized by warm temperatures, low humidity, and well-distributed rainfall. The crop thrives in temperatures ranging from 25°C to 35°C, making it well-suited for the hot and dry climates of Rajasthan, Gujarat, and Haryana.
- In terms of rainfall, cluster beans require a moderate amount of rainfall, typically between 400 mm to 600 mm per annum. The crop is relatively drought-tolerant, but consistent moisture levels are necessary for optimal growth.
- Soil-wise, cluster beans can grow in a variety of soil types, but they prefer well-drained sandy loam or loamy soils with a pH range of 7.0 to 8.0. The crop is sensitive to waterlogging, so regions with poor drainage are not suitable for cluster bean cultivation.

Overall, the agroclimatic characteristics of regions growing cluster beans are defined by warm temperatures, moderate rainfall, and well-drained soils.

4.3.2. Optimum Weather Conditions for Sowing the Crop

In India, the optimum weather conditions required for sowing cluster beans are:

- The ideal time for sowing cluster beans in India is during the monsoon season, typically in July-August. At this time, the soil has sufficient moisture, and the temperature is warm but not excessively hot.
  - Temperature-wise, the ideal temperature for sowing cluster beans ranges from 25°C to 30°C. Temperatures above 35°C can lead to poor germination and seedling establishment.

- In terms of rainfall, cluster beans require a well-distributed rainfall of at least 400 mm during the growing season. The soil should have a consistent moisture level, but excessive rainfall can lead to waterlogging and poor growth.
- Relative humidity should be moderate, ranging from 50% to 70%. High humidity can lead to fungal diseases, while low humidity can cause moisture stress.
- Wind speed should be moderate; not exceeding 15 km/h. Strong winds can damage the crop and reduce yields.

Overall, the optimum weather conditions for sowing cluster beans in India are characterized by warm temperatures, moderate rainfall, and well-distributed moisture.

4.3.3. Weather Conditions Favorable for Getting High Productivity

In India, the optimum weather conditions required for sowing cluster beans are:

- The ideal time for sowing cluster beans in India is during the monsoon season, typically in July-August. At this time, the soil has sufficient moisture, and the temperature is warm but not excessively hot.
- Temperature-wise, the ideal temperature for sowing cluster beans ranges from 25°C to 30°C. Temperatures above 35°C can lead to poor germination and seedling establishment.
- In terms of rainfall, cluster beans require a well-distributed rainfall of at least 400 mm during the growing season. The soil should have a consistent moisture level, but excessive rainfall can lead to waterlogging and poor growth.
- Relative humidity should be moderate, ranging from 50% to 70%. High humidity can lead to fungal diseases, while low humidity can cause moisture stress.
- Wind speed should be moderate; not exceeding 15 km/h. Strong winds can damage the crop and reduce yields.

Overall, the optimum weather conditions for sowing cluster beans in India are characterized by warm temperatures, moderate rainfall, and well-distributed moisture.

4.3.4. Effect of Extreme Weather Conditions

Extreme weather events can have a significant impact on cluster beans crops in India. Here are some of the effects:

- **Drought**: Cluster beans are relatively drought-tolerant, but prolonged drought can lead to reduced yields, poor pod formation, and lower quality seeds.
- **Excessive rainfall**: Heavy rainfall can cause waterlogging, which can lead to root rot, reduced growth, and lower yields.
- **Heatwaves:** High temperatures above 40°C can cause moisture stress, reduce pollination, and lead to poor pod formation.
- **Floods:** Flooding can cause complete loss of the crop, especially if the water remains stagnant for an extended period.
- **Hailstorms:** Hail can damage the crop, causing physical injury to the plants, breaking of branches, and reducing yields.
- **Cold waves**: Cluster beans are sensitive to frost, and cold waves can damage or kill the crop, especially during the flowering and pod-formation stages.
- **Windstorms**: Strong winds can damage the crop, causing lodging, breaking of branches, and reducing yields.

These extreme weather events can not only reduce yields but also affect the quality of the cluster beans crop. Therefore, it is essential for farmers to take necessary precautions, such as adopting climate-resilient varieties, using irrigation and drainage systems, and following crop insurance schemes to mitigate the risks associated with extreme weather events.

4.4. Cowpea Bean (*Vigna unguiculata*)

Cowpea beans, also known as lobia, are a type of bean widely grown in India (Fig. 4.4). They are primarily cultivated in states like Maharashtra, Gujarat, and Karnataka.

Fig. 4.4: Cowpea bean crop

The country produced over 0.3 million metric tons of cowpea beans in 2020-21. Cowpea beans are used in various dishes, including curries, stews, and salads.

- **Area:** Cowpea is grown in various parts of India, with major production areas in the states of Maharashtra, Madhya Pradesh, Uttar Pradesh, and Rajasthan. The crop is typically grown in areas with well-drained soils and a warm climate.
- **Production**: According to the Indian Council of Agricultural Research, the country produces around 1.5-2.0 million tons of cowpeas annually. The production is mostly concentrated in the kharif season, which spans from June to September.
- **Importance**: Cowpea is a vital crop in India due to its high nutritional value, drought tolerance, and versatility. The crop is an excellent source of protein, fiber, and minerals like potassium and iron. Cowpeas are consumed in various forms, including fresh, dried, and processed into value-added products like dal and flour.

In addition to its culinary uses, cowpea also plays a significant role in crop rotation and soil fertility management. As a leguminous crop, cowpea has the ability to fix atmospheric nitrogen, which benefits subsequent crops and improves soil health. Overall, cowpea is a valuable crop in India, providing nutritional security, supporting sustainable agriculture practices, and contributing to the country's food and nutritional security.

4.4.1. Agroclimatic Conditions Suitable for Growing Cowpea

Cowpea is a warm-season crop that thrives in India's tropical and subtropical regions. The crop is typically grown during the kharif season, which spans from June to September. The ideal agroclimatic conditions for growing cowpea in India are:

- **Temperature**: Cowpea requires a warm temperature, with an optimal range of 20-30°C. The crop is sensitive to frost and cannot tolerate temperatures below 10°C.
- **Soil**: Cowpea can grow well in a wide range of soils, but it prefers well-drained, sandy loam soils with a pH between 6.0 and 7.0.
- **Moisture**: Cowpea is drought-tolerant, but it requires adequate moisture for optimal growth. The crop requires around 500-600 mm of rainfall or irrigation during its growth period.
- **Sunlight**: Cowpea requires full sunlight to produce well, and it should be planted in areas that receive direct sunlight for at least 6 hours a day.

- **Elevation**: Cowpea can be grown up to an elevation of 1,500 meters above mean sea level. Region-wise, cowpea is grown in various parts of India, including the Indo-Gangetic Plain, the Deccan Plateau, and the Eastern Coastal Plains.

4.4.2. Optimum Weather Conditions Fir Sowing

In India, the ideal time for sowing cowpea is during the kharif season, which typically begins in June or July, depending on the region. The optimum weather conditions required for sowing cowpea crop in India are:

- **Temperature**: The ideal temperature for sowing cowpea is between 20-25°C. This allows for optimal germination and seedling growth.
- **Soil Temperature**: The soil temperature should be around 20-22°C for optimal germination.
- Moisture: Adequate moisture is essential for germination and seedling growth. The soil should be moist but not waterlogged.
- **Rainfall**: Cowpea requires around 500-600 mm of rainfall or irrigation during its growth period. Sowing should be done just before the onset of monsoon rains or when the soil has sufficient moisture.
- **Relative Humidity**: A relative humidity of 60-80% is ideal for cowpea growth.
- **Sunshine**: Cowpea requires full sunlight for optimal growth, and sowing should be done in areas that receive direct sunlight for at least 6 hours a day.
- **Wind**: Moderate winds are beneficial for cowpea growth, but strong winds can damage the crop.

4.4.3. Weather Conditions Favorable for Getting High Yields

The weather conditions favorable for getting high productivity of cowpea are:

- **Temperature:** A temperature range of 20-30°C is ideal for cowpea growth and development. Optimal temperatures during different growth stages are: 20-25°C during germination and seedling growth, 25-28°C during flowering and pod formation, and 20-22°C during seed filling.
- **Moisture**: Adequate and well-distributed rainfall or irrigation is essential for high cowpea productivity. The crop requires around 500-600 mm of rainfall or irrigation during its growth period.
- **Rainfall Pattern:** A rainfall pattern with a dry spell during the seed filling stage is beneficial for cowpea productivity. This helps in reducing the incidence of diseases and promotes proper seed filling.

- **Relative Humidity**: A relative humidity of 60-80% is ideal for cowpea growth and development.
- Sunshine: Cowpea requires full sunlight for optimal growth and development. A minimum of 6 hours of direct sunlight per day is essential for high productivity.
- **Soil Temperature**: Optimal soil temperature for cowpea growth is between 20-25°C.
- **Wind**: Moderate winds are beneficial for cowpea growth, but strong winds can damage the crop.

By providing these favorable weather conditions, cowpea productivity can be significantly improved, leading to higher yields and better quality seeds.

4.4.4. Effect of Extreme Weather Events

Extreme weather events have become increasingly frequent and severe in India, significantly impacting cowpea crops.

- **Drought**: Cowpea is relatively drought-tolerant, but severe droughts can still cause significant yield losses. Droughts during the flowering and pod formation stages are particularly damaging.
- **Heatwave**: High temperatures above 35°C can cause flower drop, reduce pollination, and lead to poor seed set. Heatwaves during the reproductive stage can result in significant yield losses.
- **Flooding**: Excessive rainfall and flooding can cause root rot, reduce soil oxygen, and increase the risk of diseases. Flooding during the germination and seedling stages can be particularly devastating.
- **Hailstorms**: Hailstorms can cause physical damage to cowpea plants, reducing yields and affecting quality.

Extreme weather events not only impact cowpea yields but also affect the quality of the crop. Changes in temperature and precipitation patterns can also alter the distribution and prevalence of pests and diseases, further exacerbating the challenges faced by cowpea farmers in India. To mitigate the effects of extreme weather events, Indian farmers are adopting climate-resilient agricultural practices, such as using drought-tolerant varieties, implementing conservation agriculture, and leveraging weather-based crop insurance programs.

4.5. Broad Beans (*Vicia faba*)

Broad beans, also known as fava or faba beans, are another type of bean grown in India (Fig. 4.5). They are primarily cultivated in the northern states of Punjab, Haryana, Uttar Pradesh, and Himachal Pradesh.

Fig. 4.5: Broad beans crop

The country produced over 0.2 million metric tons of broad beans in 2020-21. Broad beans are a nutritious and versatile crop, rich in protein, fiber, and other essential nutrients. They are used in various forms, including fresh, dried, and frozen.

- **Area:** Faba beans are primarily grown in the states of Madhya Pradesh, Maharashtra, Rajasthan, and Uttar Pradesh. They are typically cultivated in areas with mild winters and cool summers.
- **Production**: According to the Indian Council of Agricultural Research, the country produces around 0.5-0.6 million tons of fava beans annually. The production is mostly concentrated in the rabi season, which spans from October to March.
- **Importance:** Faba beans are an important crop in India due to their nutritional value and versatility. They are a rich source of protein, fiber, and minerals like potassium and iron. Fava beans are consumed in various forms, including fresh, dried, and processed into value-added products like dal and flour.

In addition to their culinary uses, fava beans also play a significant role in crop rotation and soil fertility management. As a leguminous crop, they have the ability to fix atmospheric nitrogen, which benefits subsequent crops and improves soil health. Overall, fava beans are a valuable crop in India, providing nutritional security and supporting sustainable agriculture practices.

4.5.1. Agroclimatic conditions of the regions growing faba beans

The faba beans growing regions in India are characterized by distinct agroclimatic conditions that support the growth and development of crop.

- **Temperature**: Fava beans are grown in regions with mild winters and cool summers. The ideal temperature for fava bean cultivation ranges from 15°C to 25°C. Madhya Pradesh, Maharashtra, and Rajasthan experience a temperate climate with cold winters and warm summers, making them suitable for fava bean production.

- **Soil**: Fava beans can thrive in a wide range of soils, but well-drained loamy soils with a pH between 6.0 and 7.0 are considered ideal. The alluvial soils of the Indo-Gangetic Plain and the black cotton soils of Madhya Pradesh and Maharashtra provide a suitable environment for fava bean cultivation.
- **Moisture**: Fava beans require adequate moisture, especially during the flowering and pod formation stages. The regions with fava bean cultivation in India receive an average annual rainfall of 600-1000 mm, which is sufficient to support crop growth. Irrigation facilities are also available in these regions to supplement rainfall during periods of drought.
- **Elevation**: Fava beans are grown in regions with varying elevations, ranging from 100 to 1500 meters above mean sea level. The crop is cultivated in the plains of Madhya Pradesh, Maharashtra, and Uttar Pradesh, as well as in the foothills of the Himalayas.

Overall, the agroclimatic conditions in the fava beans growing regions of India provide a favorable environment for the crop's growth and development, enabling the country to produce a significant quantity of fava beans annually.

4.5.2. Optimum Period Of Sowing and Support Requirement for Creepers

Faba beans require specific weather conditions for optimal sowing and growth. The ideal time for sowing fava beans in India is from October to November, when the weather is cool and the soil has sufficient moisture.

- The crop requires a period of cool weather during the germination and seedling stages, with temperatures ranging from 10°C to 20°C.
- As the plants grow, they require support to prevent lodging and promote upright growth. Fava bean plants can grow quite tall, up to 1-2 meters, and have a tendency to fall over in the wind. To provide support, farmers typically use a system of stakes and twine or a trellis.
- The support system should be installed when the plants are around 30-40 cm tall, and the twine or trellis should be adjusted as the plants grow to keep them upright and secure. This support system helps to promote healthy growth, prevent damage from wind and rain, and make harvesting easier.

4.5.3: Weather Conditions Favorable for High Productivity

Faba beans require a specific set of weather conditions to achieve high productivity. Here are the favorable weather conditions:

- **Cool winters**: Fava beans thrive in cool winters with temperatures ranging from 10°C to 20°C. This allows for optimal growth and development during the vegetative stage.
- **Mild springs**: A mild spring with temperatures between 15°C to 25°C promotes healthy flowering and pod formation.
- **Adequate moisture**: Fava beans require consistent and adequate moisture, especially during the flowering and pod formation stages. An annual rainfall of 600-1000 mm is considered ideal.
- **Low humidity**: Low humidity during the reproductive stage helps prevent fungal diseases that thrive in humid environments.
- **Moderate sunshine:** Fava beans require moderate sunshine, especially during the pod formation stage. Direct sunlight promotes healthy growth and development.
- **Absence of extreme weather events**: Fava beans are sensitive to extreme weather events like heavy rainfall, hailstorms, and strong winds. The absence of such events promotes healthy growth and reduces the risk of damage to the crop.

Overall, a cool and relatively dry winter, followed by a mild spring with adequate moisture, is considered ideal for high productivity of fava beans.

4.5.4. Effect of Extreme Weather Events

Extreme weather events have taken a devastating toll on faba bean crops in India. Rising temperatures, erratic rainfall patterns, and increased frequency of heatwaves and droughts have significantly impacted fava bean production.

- **Temperature Extremes:** Fava beans are sensitive to extreme temperatures. Heatwaves can cause damage to the crop, especially during the flowering and pod formation stages. On the other hand, cold temperatures can also impact fava bean yields.
- **Rainfall Variability**: Erratic rainfall patterns can lead to waterlogged soil conditions, causing root rot and reducing fava bean yields. Conversely, droughts can also impact fava bean production, especially during the critical growth stages.
- **Regional Impacts**: Different regions in India are experiencing varying levels of extreme weather events. For example, Rajasthan and

Maharashtra have faced severe heatwaves, while Kerala has experienced devastating floods.

- **Economic Consequences:** The economic impacts of extreme weather events on fava bean production are substantial. The damage to crops, infrastructure, and human livelihoods can have long-lasting effects on the agricultural sector.

To mitigate the effects of extreme weather events on fava bean production, Indian policymakers are taking steps to enhance disaster preparedness, improve early warning systems, and promote climate-resilient agricultural practices.

4.6. Peas (*Pisum sativum*)

Peas are an important legume crop in India (Fig. 4.6), widely cultivated for their nutritional and economic benefits.

Fig. 4.6: Peas crop

- **Area**: Peas are grown in various parts of India, with major production areas in the states of Uttar Pradesh, Madhya Pradesh, Maharashtra, and Punjab. According to the Indian Council of Agricultural Research, the total area under pea cultivation in India is around 2.5-3.0 million hectares.
- **Production**: India is one of the largest producers of peas in the world. The country produces around 4-5 million tons of peas annually, with an average yield of around 1.5-2.0 tons per hectare.
- **Importance**: Peas are a vital crop in India due to their high nutritional value, ease of cultivation, and versatility. They are an excellent source of protein, fiber, and minerals like potassium and iron. Peas are consumed in various forms, including fresh, dried, and processed into value-added products like dal and flour.
- In addition to their culinary uses, peas also play a significant role in crop rotation and soil fertility management. As a leguminous crop, peas have

the ability to fix atmospheric nitrogen, which benefits subsequent crops and improves soil health.

Overall, peas are a valuable crop in India, providing nutritional security, supporting sustainable agriculture practices, and contributing to the country's food and nutritional security.

4.6.1. Agroclimatic conditions required for growing peas

Peas are a cool-season crop that thrives in India's temperate and subtropical regions. In India, peas are often grown as a winter crop, taking advantage of the cool temperatures and moisture available during this period. The ideal agroclimatic conditions for growing peas in India are:

- **Temperature**: Peas require a cool temperature, with an optimal range of 10-20°C. The crop is sensitive to high temperatures and cannot tolerate temperatures above 25°C.
- **Soil:** Peas can grow well in a wide range of soils, but they prefer well-drained, sandy loam soils with a pH between 6.0 and 7.0.
- **Moisture**: Peas require adequate moisture, especially during the germination and flowering stages. The crop requires around 400-500 mm of rainfall or irrigation during its growth period.
- **Sunlight**: Peas require partial shade to full sun, depending on the temperature. In warmer climates, peas prefer partial shade, while in cooler climates, they can tolerate full sun.
- **Elevation:** Peas can be grown up to an elevation of 2,000 meters above mean sea level.
- **Region-wise:** Peas are grown in various parts of India, including the Indo-Gangetic Plain, the Deccan Plateau, and the Eastern Coastal Plains. The crop is typically grown during the rabi season, which spans from October to March.

4.6.2. Optimum Weather Conditions Required for Sowing Peas

Peas are a cool-season crop that thrives in India's temperate and subtropical regions. The ideal agroclimatic conditions for growing peas in India are:

- **Temperature**: Peas require a cool temperature, with an optimal range of 10-20°C. The crop is sensitive to high temperatures and cannot tolerate temperatures above 25°C.
- **Soil:** Peas can grow well in a wide range of soils, but they prefer well-drained, sandy loam soils with a pH between 6.0 and 7.0.

- **Moisture**: Peas require adequate moisture, especially during the germination and flowering stages. The crop requires around 400-500 mm of rainfall or irrigation during its growth period.
- **Sunlight:** Peas require partial shade to full sun, depending on the temperature. In warmer climates, peas prefer partial shade, while in cooler climates, they can tolerate full sun.
- **Elevation:** Peas can be grown up to an elevation of 2,000 meters above mean sea level.
- **Region-wise**: Peas are grown in various parts of India, including the Indo-Gangetic Plain, the Deccan Plateau, and the Eastern Coastal Plains. The crop is typically grown during the rabi season, which spans from October to March.

4.6.3. Weather Conditions Favorable Fir Getting High Productivity

The weather conditions favorable for getting high productivity of peas are:

- **Temperature**: A temperature range of 10-20°C is ideal for pea growth and development. Optimal temperatures during different growth stages are: 10-15°C during germination and seedling growth, 15-18°C during flowering and pod formation, and 10-12°C during seed filling.
- **Moisture**: Adequate and well-distributed rainfall or irrigation is essential for high pea productivity. The crop requires around 400-500 mm of rainfall or irrigation during its growth period.
- **Rainfall Pattern**: A rainfall pattern with a dry spell during the seed filling stage is beneficial for pea productivity. This helps in reducing the incidence of diseases and promotes proper seed filling.
- **Relative Humidity**: A relative humidity of 60-80% is ideal for pea growth and development.
- **Sunshine**: Peas require partial shade to full sun, depending on the temperature. In warmer climates, peas prefer partial shade, while in cooler climates, they can tolerate full sun.
- **Soil Moisture**: Adequate soil moisture is essential for pea growth, especially during the germination, flowering, and pod formation stages.

By providing these favorable weather conditions, pea productivity can be significantly improved, leading to higher yields and better quality seeds.

4.6.4: Effect of extreme weather events

Extreme weather events have become increasingly frequent and severe in India, significantly impacting pea crops. Here's how:

- **Droughts**: Peas are sensitive to drought, especially during the germination and flowering stages. Droughts can lead to reduced yields, poor seed quality, and increased susceptibility to pests and diseases.
- **Heatwaves**: High temperatures above 25°C can cause flower drop, reduce pollination, and lead to poor seed set. Heatwaves during the reproductive stage can result in significant yield losses.
- **Flooding**: Excessive rainfall and flooding can cause root rot, reduce soil oxygen, and increase the risk of diseases. Flooding during the germination and seedling stages can be particularly devastating.
- **Hailstorms**: Hailstorms can cause physical damage to pea plants, reducing yields and affecting quality.
- Extreme weather events not only impact pea yields but also affect the quality of the crop. Changes in temperature and precipitation patterns can also alter the distribution and prevalence of pests and diseases, further exacerbating the challenges faced by pea farmers in India.

To mitigate the effects of extreme weather events, Indian farmers are adopting climate-resilient agricultural practices, such as using drought-tolerant varieties, implementing conservation agriculture, and leveraging weather-based crop insurance programs.

4.7. Conclusion

Agroclimatology plays a vital role in understanding the complex relationships between climate, weather, and bean crop growth. Different types of beans, such as kidney beans, French beans, cluster beans, peas and cowpeas, have unique agroclimatic requirements that must be met to ensure optimal growth and productivity. Understanding these requirements enables farmers and agricultural planners to make informed decisions about crop selection, planting dates, and management practices, ultimately leading to improved crop yields, reduced risks, and enhanced food security. By applying agroclimatic principles, bean farmers in India can optimize their crop production, improve their livelihoods, and contribute to the country's food and nutritional security.

5

Agroclimatology of Root and Tuber Type Vegetables

Root-type vegetables are a diverse group of underground plant structures that are edible and serve as a vital source of nutrition for humans. These vegetables can be further categorized into two main types: true roots, such as carrots, beets, turnips, and radishes, which are modified taproots that store food; and tuberous roots, such as Potatoes, sweet potatoes and yams, which are modified stems that store food. A root is a compact, often enlarged storage organ with hairy stems that develops from root tissue. So a tuber is a root crop, but a plant can be a root and not a tuber. Carrots and cassava are root vegetable crops. Potatoes, sweet potatoes and yams, on the other hand, are edible tuber crops (Fig. 5.1).

Fig. 5.1: Root and tuber type vegetables

The primary difference in propagation lies in the plant part used: tubers, like potatoes, are propagated from modified stems or roots with "eyes" (nodes) that sprout, while root crops, like carrots, are propagated from seeds or root cuttings, as their roots don't have these propagation points. Another major difference is that tubers will have long shelf life compared to roots.

- **Tubers**
  - **Origin**: Tubers are modified stems or roots that store nutrients.
  - **Propagation**: They can be propagated vegetatively by using pieces of the tuber that contain "eyes" (nodes) which sprout and grow into new plants.
  - **Examples**: Potatoes, sweet potatoes, and cassava.
- **Root Crops**
  - **Origin**: Root crops are enlarged taproots that store nutrients.
  - **Propagation:** They are typically propagated using seeds or root cuttings, as their roots don't have the ability to sprout like tubers.
  - **Examples**: Carrots, beets, radishes, and turnips.
- In India both tuber and root-type vegetables are a significant component of the country's vegetable production, playing a crucial role in the nation's food security and economy. India is one of the largest producers of root-type vegetables in the world, with a total production of over 15 million tonnes annually. The major root vegetable-producing states in India include Uttar Pradesh, Punjab, Bihar, Gujarat, and Madhya Pradesh. Root vegetables are grown in almost all parts of the country, with different regions specializing in different types of root vegetables. These vegetables are not only an essential part of Indian cuisine but also provide livelihood opportunities for millions of farmers, traders, and laborers involved in their production, processing, and marketing.
- The importance of root-type vegetables in India extends beyond their nutritional value and economic significance. They are also an integral part of the country's cultural heritage, with different regions having their unique recipes and traditions associated with these vegetables. This chapter provides an overview of the agroclimatology of root-type vegetables in India, including their area, production, and importance, as well as the various factors that influence their growth and productivity.
- The importance of root-type vegetables in India extends beyond their nutritional value and economic significance. They are also an integral part of the country's cultural heritage, with different regions having their unique recipes and traditions associated with these vegetables. This chapter provides an overview of the agroclimatology of root-type vegetables in India, including their area, production, and importance, as well as the various factors that influence their growth and productivity.

5.1. Potatoes (*Solanum tuberosum*)

Potatoes are one of the most versatile and widely consumed staple crop globally, playing a vital role in food security and economies worldwide, including India (Fig. 5.2).

Fig. 5.2: Potato crop

They are a type of underground plant part that is often classified as both a tuber and a root crop, due to its characteristics. As a tuber, potatoes are a type of swollen stem that grows underground, producing multiple stems and roots. On the other hand, they are also considered a root crop due to their edible root system.

- In terms of area, potatoes are grown in over 130 countries worldwide, with India being one of the largest producers. According to the Food and Agriculture Organization (FAO) of the United Nations, India is the second-largest potato producer in the world, accounting for around 25% of the global production.
- In India, potatoes are grown in various states, with Uttar Pradesh, Punjab, and Bihar being the top three potato-producing states. The crop is typically sown in the winter season, from October to December, and harvested between February and April.
- India's potato production has been steadily increasing over the years, thanks to improved farming practices, better crop varieties, and favorable government policies. According to the Indian Council of Agricultural Research (ICAR), the country's potato production has grown from around 25 million tons in 2000-01 to over 50 million tons in 2020-21.
- Potatoes play a vital role in India's food security and economy. They are a staple food for millions of people, particularly in rural areas, and are used in a variety of dishes, from curries and fries to salads and snacks. The crop is also an important source of income for farmers, with many relying on potato cultivation as their primary source of livelihood.

- In addition to its food and economic importance, potatoes also have significant nutritional value. They are rich in complex carbohydrates, fiber, and antioxidants, making them an excellent addition to a healthy diet.

Overall, potatoes are a vital crop in India, providing food, income, and nutrition to millions of people. The government and agricultural organizations are working to further improve potato production and productivity, which will help to ensure the country's food security and support the livelihoods of farmers.

5.1.1. Agroclimatic Requirements for Growing Potatoes in India

Potatoes are a temperate crop that requires a specific set of agroclimatic conditions to grow. In India, potatoes are typically grown in the winter season, and the agroclimatic requirements for growing potatoes vary across different regions. Here are the ideal agroclimatic requirements for growing potatoes in India:

- **Temperature**: Potatoes require a cool temperature to grow, with optimal temperatures ranging from 15°C to 20°C. Temperatures above 25°C can lead to reduced yields and poor tuber quality.
- **Moisture:** Potatoes require adequate moisture, especially during the tuber formation stage. The ideal rainfall for potato cultivation is around 600-800 mm, with supplemental irrigation provided as needed.
- **Soil**: Potatoes can grow in a wide range of soils, but well-drained, loose soils with a pH between 4.5 and 7.0 are ideal.
- **Day Length**: Potatoes require a short day length to induce tuberization. In India, the winter season provides the ideal day length for potato cultivation.
- **Altitude**: Potatoes can be grown at a wide range of altitudes, from sea level to over 2,000 meters. However, the optimal altitude for potato cultivation in India is between 1,000 and 1,500 meters.

Regions with ideal agroclimatic conditions for potato cultivation in India include:

- The Indo-Gangetic Plains, which include states such as Uttar Pradesh, Punjab, and Bihar.
- The Himalayan foothills, which include states such as Himachal Pradesh, Uttarakhand, and Jammu and Kashmir.
- The Nilgiri Hills in Tamil Nadu.

Overall, the agroclimatic requirements for growing potatoes in India are specific and vary across different regions. By selecting the right variety and growing potatoes in the ideal agroclimatic conditions, farmers can achieve high yields and good tuber quality.

5.1.2. Optimum Weather Conditions for Planting Potatoes

Potatoes are primarily propagated through seed tubers, which are small, whole potatoes or pieces of potatoes that are used to grow new plants. The seed tubers are obtained from certified disease-free seed potato production areas or from the farmer's own crop.

- In India, potatoes are typically planted in the winter season, from October to December, depending on the region and climate. The optimum conditions required for planting potatoes in India include a cool temperature, adequate moisture, and well-draining soil.
- The ideal temperature for planting potatoes is between 10°C and 20°C. Temperatures above 25°C can lead to poor germination and reduced yields. The soil should be well-draining and loose, with a pH between 4.5 and 7.0.
- Before planting, the seed tubers are usually treated with a fungicide to prevent disease. The tubers are then cut into pieces, each containing at least one "eye" or bud. The cut pieces are allowed to heal for a day or two before planting to reduce the risk of rot.
- In India, potatoes are often planted in ridges or beds, which help to improve drainage and soil aeration. The seed pieces are planted at a depth of around 5-7 cm, with the "eyes" facing upwards. The soil is then covered with a layer of organic mulch or straw to retain moisture and suppress weeds.
- After planting, the crop requires adequate moisture, especially during the germination and tuber formation stages. Supplemental irrigation is often provided, especially in areas with low rainfall.

5.1.3. Ideal Weather Conditions Required for High Productivity of Potatoes

To achieve high productivity of potatoes, specific weather conditions are necessary. During the germination and emergence stage, potatoes require a cool and moist climate.

- Temperatures between 10°C and 15°C are ideal for germination, while temperatures above 20°C can lead to poor germination and reduced yields.

- After emergence, potatoes require a moderate temperature, ranging from 15°C to 20°C, to promote healthy growth and tuber formation. Temperatures above 25°C can lead to stress, reducing yields and affecting tuber quality.
- Adequate moisture is also essential for high potato productivity. Potatoes require consistent and well-distributed rainfall or irrigation throughout the growing season. A total rainfall of around 600-800 mm is considered ideal for potato cultivation.
- Relative humidity also plays a crucial role in potato productivity. A relative humidity of 60-80% is considered ideal, as it helps to prevent moisture stress and promotes healthy growth.
- Wind protection is also necessary, as strong winds can damage the crop and reduce yields. In areas with high wind speeds, windbreaks or row covers can be used to protect the crop.
- Finally, potatoes require a certain amount of sunshine to produce a good yield. Bright, sunny weather with around 6-8 hours of direct sunlight per day is ideal for potato cultivation.

5.1.4. Effect of Extreme Weather Events on Potato Crop

To achieve high productivity of potatoes, specific weather conditions are necessary. During the germination and emergence stage, potatoes require a cool and moist climate.

- Temperatures between 10°C and 15°C are ideal for germination, while temperatures above 20°C can lead to poor germination and reduced yields.After emergence, potatoes require a moderate temperature, ranging from 15°C to 20°C, to promote healthy growth and tuber formation. Temperatures above 25°C can lead to stress, reducing yields and affecting tuber quality.
- Adequate moisture is also essential for high potato productivity. Potatoes require consistent and well-distributed rainfall or irrigation throughout the growing season. A total rainfall of around 600-800 mm is considered ideal for potato cultivation.
- Relative humidity also plays a crucial role in potato productivity. A relative humidity of 60-80% is considered ideal, as it helps to prevent moisture stress and promotes healthy growth.
- Wind protection is also necessary, as strong winds can damage the crop and reduce yields. In areas with high wind speeds, windbreaks or row covers can be used to protect the crop.

- Finally, potatoes require a certain amount of sunshine to produce a good yield. Bright, sunny weather with around 6-8 hours of direct sunlight per day is ideal for potato cultivation.

5.1.5. Weather conditions suitable for for storage to extend shelf life

To extend the shelf life of potatoes, they need to be stored in a cool, dark, and well-ventilated place with specific weather conditions. The ideal storage conditions for potatoes are:

- **Temperature**: The storage temperature should be between 4°C and 10°C. Temperatures above 10°C can lead to sprouting and rot, while temperatures below 4°C can cause the potatoes to become sweet.
- **Humidity:** The relative humidity should be around 80-90%. High humidity can lead to rot and mold, while low humidity can cause the potatoes to become dehydrated.
- **Light**: Potatoes should be stored in complete darkness. Light can stimulate sprouting and chlorophyll production, which can lead to a decrease in quality.
- **Ventilation**: Good ventilation is essential for maintaining the quality of stored potatoes. A consistent flow of air helps to remove moisture and heat, reducing the risk of rot and sprouting.
- **Air circulation**: Air circulation is also crucial for maintaining the quality of stored potatoes. A gentle circulation of air helps to prevent the buildup of moisture and heat.
- **Carbon dioxide**: A low concentration of carbon dioxide is beneficial for storing potatoes. High levels of carbon dioxide can lead to a buildup of ethylene, which can stimulate sprouting.

The ideal storage conditions for potatoes are cool, dark, and well-ventilated, with consistent temperature, humidity, and air circulation. By maintaining these conditions, the shelf life of potatoes can be extended, and their quality can be maintained.

5.2. Sweet Potato (*Ipomoea batatas*)

Sweet potato is an important root crop in India, providing a rich source of nutrients, particularly vitamins A and C, potassium, and fiber (Fig. 5.3). Here's an overview of sweet potato cultivation in India.

Fig. 5.3: Sweet potato crop

- **Area and Production**: Sweet potato is grown in many parts of India, with major producing states including Uttar Pradesh, Bihar, Odisha, and Tamil Nadu. According to the Indian Council of Agricultural Research (ICAR), the country produces around 1.5-2 million tons of sweet potatoes annually, with an average yield of 20-25 tons per hectare.
- **Importance:** Sweet potato is a vital crop in India, especially for small and marginal farmers, due to its adaptability to poor soil conditions, drought tolerance, and short growth period. It serves as a food security crop, providing a reliable source of income and nutrition for rural communities.

In addition, sweet potato has several industrial applications, such as starch production, ethanol generation, and animal feed. Research institutions and organizations are working to improve sweet potato yields, disease resistance, and nutritional content, further enhancing its importance in Indian agriculture.

5.2.1. Agroclimatic Conditions Required for Growing Sweet Potato

Sweet potatoes are a tropical crop that thrives in India's diverse agroclimatic conditions. Here are the ideal conditions required to grow sweet potatoes in India:

- **Temperature**: Sweet potatoes require a warm temperature, typically between 20°C to 30°C, for optimal growth. Temperatures above 35°C can lead to sun scorching, while temperatures below 10°C can cause chilling injury.
- **Moisture:** Sweet potatoes need adequate moisture, especially during the germination and tuberization stages. A well-distributed annual rainfall of 750-1,000 mm is ideal. However, sweet potatoes can tolerate some drought, making them suitable for areas with limited irrigation facilities.
- **Soil**: Sweet potatoes can grow in a wide range of soils, but they prefer well-draining, loose soils with a pH between 5.5 and 6.5.

- **Sunlight**: Sweet potatoes require full sun to partial shade. They thrive in areas receiving direct sunlight for at least 6 hours a day.
- **Elevation:** Sweet potatoes can be grown up to an elevation of 1,500 meters above sea level.
- **Seasonality:** In India, sweet potatoes are typically grown during the kharif (June-October) and rabi (November-February) seasons. However, the exact sowing time may vary depending on the region and climate.
- **Regional Variations**: The ideal conditions for growing sweet potatoes vary across different regions in India. For instance, in the northern plains, sweet potatoes are grown during the kharif season, while in the southern states, they are grown during the rabi season. In the northeastern states, sweet potatoes are grown during the spring-summer season (March-July), and in the coastal regions, they are grown during the post-monsoon season (October-December).

5.2.2. Optimum Weather Conditions for Planting Sweet Potato

The optimum weather conditions required for planting sweet potatoes are:

- **Temperature**: A warm temperature is essential for sweet potato growth. The ideal temperature for planting sweet potatoes ranges from 20°C to 30°C. Temperatures above 35°C can lead to sun scorching, while temperatures below 10°C can cause chilling injury.
- **Rainfall:** Sweet potatoes require adequate moisture, especially during the germination and tuberization stages. A well-distributed annual rainfall of 750-1,000 mm is ideal. However, sweet potatoes can tolerate some drought, making them suitable for areas with limited irrigation facilities.
- **Soil Temperature**: The ideal soil temperature for planting sweet potatoes is between 18°C and 24°C.
- **Relative Humidity**: A relative humidity of 60-80% is suitable for sweet potato growth.
- **Wind**: Sweet potatoes prefer a wind-free environment. Strong winds can damage the plants and reduce yields.
- **Sunlight**: Sweet potatoes require full sun to partial shade. They thrive in areas receiving direct sunlight for at least 6 hours a day.
- **Planting Time**: In India, the ideal time for planting sweet potatoes varies depending on the region. In general, sweet potatoes are planted during the kharif (June-October) and rabi (November-February) seasons.

5.2.3. Ideal Weather Conditions Required to Get High Productivity

To achieve high productivity of sweet potatoes, the following ideal weather conditions are necessary:

- **Temperature:** A consistent temperature between 20°C and 25°C is ideal for sweet potato growth. Temperatures above 30°C can lead to reduced yields, while temperatures below 15°C can slow down growth.
- **Rainfall:** A well-distributed annual rainfall of 1,000-1,200 mm is ideal for sweet potato cultivation. Adequate moisture during the germination and tuberization stages is crucial for high yields.
- **Soil Moisture:** Adequate soil moisture is essential for sweet potato growth. The soil should be kept consistently moist, but not waterlogged.
- **Relative Humidity**: A relative humidity of 60-80% is suitable for sweet potato growth.
- **Sunlight**: Sweet potatoes require full sun to partial shade. They thrive in areas receiving direct sunlight for at least 6 hours a day.
- **Wind Protection**: Sweet potatoes prefer a wind-free environment. Strong winds can damage the plants and reduce yields.
- **Frost-Free Period**: A frost-free period of at least 120 days is necessary for sweet potato cultivation.

Ideal Weather Conditions During Different Growth Stages

- Germination stage (0-10 days): Consistent temperature, adequate moisture, and relative humidity.
- Vine establishment stage (10-30 days): Warm temperature, adequate moisture, and full sun.
- Tuberization stage (30-60 days): Consistent temperature, adequate moisture, and reduced sunlight.
- Maturation stage (60-120 days): Warm temperature, reduced moisture, and full sun.

5.2.4. Effect of Extreme Weather Events

Extreme weather events can have a significant impact on sweet potato crops, affecting their yield, quality, and overall sustainability. Here are some of the ways extreme weather events can impact sweet potato crops:

- Drought can lead to reduced tuber formation, smaller tuber size, and lower yields. Sweet potatoes require consistent moisture, especially during the tuber formation stage. Prolonged drought can cause stress to the plants, reducing their ability to produce tubers.

- Heavy rainfall and flooding can cause root rot, reduced yields, and decreased quality. Excess moisture can lead to an increase in disease incidence, such as root rot and leaf spot. Additionally, flooding can cause the tubers to become waterlogged, reducing their quality and market value.
- Heatwaves can cause sunscald, reduced yields, and lower quality tubers. High temperatures can cause the tubers to become misshapen, and the skin to become rough and unmarketable. Heat stress can also reduce the plants' ability to undergo tuberization.
- Cold temperatures can cause chilling injury, reduced yields, and lower quality tubers. Sweet potatoes are sensitive to cold temperatures, and exposure to temperatures below 10°C (50°F) can cause chilling injury. This can lead to a range of symptoms, including surface lesions, discoloration, and reduced quality.
- Windstorms can cause physical damage to the plants, reducing yields and affecting quality. Strong winds can cause the plants to become lodged, reducing their ability to undergo photosynthesis and produce tubers.

Extreme weather events can have a significant impact on sweet potato crops, affecting their yield, quality, and sustainability. It is essential for farmers to adopt climate-resilient agricultural practices, such as using drought-tolerant varieties, implementing irrigation management systems, and practicing conservation agriculture.

5.3. Yam (*Dioscorea species*)

Yams are a significant crop in India, particularly the elephant foot yam variety (Fig. 5.4). Here are some key details about yam cultivation in India:

- **Area and Production**: India produces around 1.40 million tons of yams annually, with an average productivity of 11.44 tons per hectare [1]. The major yam-producing states in India include Odisha, Kerala, West Bengal, and Uttar Pradesh.
- **Importance**: Yams are a vital crop in India due to their nutritional value, adaptability, and economic significance. They are rich in protein, carbohydrates, and omega-3 fatty acids, making them a valuable food source. Yams are also used in various industrial applications, such as starch production and animal feed.

Fig. 5.4: Yam crop

- **Cultivation**: Yams are typically grown in tropical and subtropical regions of India, requiring a warm and humid climate. They thrive in well-draining, fertile soils with a pH range of 5.5 to 7.2. The crop is usually planted in the summer or rainy season, and harvesting takes place around 7-9 months after planting.
- **Regional Variations**: Different regions in India cultivate yams in varying degrees. For instance, the southern parts of India, particularly Kerala and Tamil Nadu, are significant producers of elephant foot yam. In contrast, the northeastern region of India is known for cultivating other types of yams, such as the yam bean.

5.3.1. Agroclimatic Requirements for Cultivation of Yam

Yams are a tropical crop that thrives in India's diverse agroclimatic conditions. Here are the ideal agroclimatic requirements for growing yams in India:

- **Temperature**: Yams require a warm temperature, typically between 25°C and 30°C, for optimal growth. Temperatures above 35°C can lead to heat stress, while temperatures below 15°C can slow down growth.
- **Rainfall:** Yams need adequate rainfall, typically between 1,000 mm and 1,500 mm per annum, for optimal growth. Well-distributed rainfall during the growing season is essential.
- **Soil:** Yams prefer well-draining, fertile soils with a pH range of 5.5 to 7.2. They can grow in a variety of soils, including clay loams, sandy loams, and alluvial soils.
- **Sunlight**: Yams require partial shade to full sun. They thrive in areas receiving direct sunlight for at least 4-6 hours a day.
- **Elevation**: Yams can be grown up to an elevation of 1,500 meters above sea level.

- **Relative Humidity**: A relative humidity of 60-80% is suitable for yam growth.
- **Wind Protection**: Yams prefer a wind-free environment. Strong winds can damage the plants and reduce yields.
- **Seasonality**: In India, yams are typically grown during the kharif (June-October) and rabi (November-February) seasons. However, the exact sowing time may vary depending on the region and climate.
- **Regional Variations:** The ideal agroclimatic requirements for growing yams vary across different regions in India. For instance:
  - In the southern states, yams are grown during the kharif season.
  - In the northeastern states, yams are grown during the spring-summer season (March-July).
  - In the coastal regions, yams are grown during the post-monsoon season (October-December).

5.3.2. Optimum Weather Conditions Required for Planting

The optimum weather conditions required for planting yams in India are:

- **Temperature:** A warm temperature, typically between 25°C and 30°C, is essential for yam growth. Temperatures above 35°C can lead to heat stress, while temperatures below 15°C can slow down growth.
- **Rainfall**: Yams require adequate rainfall, typically between 1,000 mm and 1,500 mm per annum, for optimal growth. Well-distributed rainfall during the growing season is essential.
- **Soil Temperature**: The ideal soil temperature for planting yams is between 20°C and 25°C.
- **Relative Humidity:** A relative humidity of 60-80% is suitable for yam growth.
- **Sunlight**: Yams require partial shade to full sun. They thrive in areas receiving direct sunlight for at least 4-6 hours a day.
- **Wind**: Yams prefer a wind-free environment. Strong winds can damage the plants and reduce yields.
- **Planting Time**: In India, the ideal time for planting yams varies depending on the region. In general, yams are planted during the kharif (June-October) and rabi (November-February) seasons.
- **Regional Variations**: The ideal weather conditions for planting yams vary across different regions in India. For instance, in the southern states, yams are planted during the kharif season, while in the northeastern states, yams are planted during the spring-summer season (March-July).

5.3.3. Weather Conditions Favorable for High Productivity

To achieve high productivity in yam cultivation, the following weather conditions are favorable:

- **Temperature**: A consistent temperature between 25°C and 28°C is ideal for yam growth. This temperature range promotes healthy vine growth, tuber formation, and starch accumulation.
- **Rainfall:** A well-distributed annual rainfall of 1,200-1,500 mm is ideal for yam cultivation. Adequate rainfall during the growing season ensures optimal growth, while excessive rainfall can lead to waterlogged soil and reduced yields.
- **Soil Moisture**: Adequate soil moisture is essential for yam growth. The soil should be kept consistently moist, but not waterlogged.
- **Relative Humidity**: A relative humidity of 70-80% is favorable for yam growth.
- **Sunlight**: Yams require partial shade to full sun. They thrive in areas receiving direct sunlight for at least 4-6 hours a day.
- **Wind Protection**: Yams prefer a wind-free environment. Strong winds can damage the plants and reduce yields.
- **Frost-Free Period**: A frost-free period of at least 9 months is necessary for yam cultivation.
- **Regional Variations**: While these weather conditions are generally favorable for yam cultivation, ¡regional variations may occur. For example, in regions with high temperatures, yam cultivation may require more frequent irrigation to prevent moisture stress.

5.3.4. Effect of Extreme Weather Events

Extreme weather conditions can have a significant impact on yam crops. Here are some of the effects:

- Prolonged drought can lead to reduced tuber formation, lower yields, and decreased quality. Drought stress can also make yams more susceptible to disease and pests.
- On the other hand, excessive rainfall can cause tuber rot, leading to significant yield losses. Flooding can also reduce soil oxygen levels, further exacerbating tuber rot.
- High temperatures can cause sun scorch, leading to damaged tubers and reduced yields. Temperature fluctuations can also disrupt dormancy patterns, affecting tuber sprouting and growth.

- Cold temperatures can cause chilling injury, leading to damaged tubers and reduced yields.
- Windstorms can cause physical damage to yam plants, reducing yields and affecting tuber quality.

Extreme weather conditions can significantly impact yam crops, leading to reduced yields, lower quality tubers, and decreased farmer incomes

5.4. Cassava (*Manihot esculenta*)

Cassava, also known as tapioca, is a root crop that is grown in various parts of India (Fig. 5.5). The area under cassava cultivation in India is around 0.2 million hectares, with the majority of it being cultivated in the states of Tamil Nadu, Kerala, and Andhra Pradesh.

- In terms of production, India produces around 6-7 million tonnes of cassava annually, with an average yield of around 30-40 tonnes per hectare. Cassava is an important crop in India, particularly in the southern states where it is grown as a cash crop.
- Cassava is a drought-tolerant crop that can be grown in poor soil conditions, making it an attractive option for farmers in areas with limited irrigation facilities. The crop is also a good source of employment for rural laborers, particularly during the harvesting and processing seasons.

Fig. 5.5: Cassava crop

- The importance of cassava in India can be gauged from the fact that it is used as a raw material for the production of various products, including starch, sago, and ethanol. Cassava starch is used in the production of paper, textiles, and pharmaceuticals, while sago is used as a food ingredient. Ethanol produced from cassava is used as a biofuel, which can help reduce the country's dependence on fossil fuels.
- In addition to its industrial uses, cassava is also an important food crop in India, particularly in the southern states where it is consumed in various

forms, including as a boiled root, a fried snack, or as a flour used to make baked goods.

Cassava is an important crop in India, providing employment, income, and food security to millions of people, particularly in the rural areas.

5.4.1. Agroclimatic Conditions Required for Cassava Cultivation

Cassava is a tropical crop that requires a warm and humid climate to grow. In India, cassava is typically grown in regions with high temperatures and adequate rainfall.

- The ideal temperature for cassava cultivation ranges from 25°C to 30°C, although it can tolerate temperatures up to 35°C.
- Cassava requires a well-distributed annual rainfall of around 500-750 mm, with a minimum of 300 mm during the growing season. However, it can also be grown under irrigation, particularly in areas with limited rainfall.
- In terms of soil, cassava can be grown on a wide range of soils, including sandy loams, clay loams, and laterites. However, it prefers well-draining, acidic soils with a pH range of 5.5-6.5.
- Cassava is typically planted at the beginning of the rainy season, and it takes around 6-12 months to mature, depending on the variety and growing conditions. The crop requires full sun to partial shade, and it can be grown at elevations up to 1,000 meters above sea level.
- In India, the major cassava-growing regions include the southern states of Tamil Nadu, Kerala, and Andhra Pradesh, where the climate is warm and humid, and the rainfall is adequate. These regions provide ideal conditions for cassava cultivation, and they account for the majority of the country's cassava production. Cassava planting requires optimum weather conditions to ensure proper germination and establishment.

5.4.2. Ideal Weather Conditions for Planting Cassava

The ideal weather conditions for planting cassava include:

- **Warm temperatures**, with soil temperatures ranging from 25°C to 30°C. This warm temperature range promotes rapid germination and seedling growth.
- **Adequate moisture**, with a minimum of 300 mm of rainfall during the growing season. Cassava requires consistent moisture, especially during the first few weeks after planting.
- **High humidity**, with an average relative humidity of 60-80%. This helps to maintain soil moisture and promote healthy seedling growth.

- **Low wind** speeds, as high winds can damage young seedlings and disrupt soil moisture.
- **Partial shade** to full sun, depending on the variety and growing conditions. Cassava can tolerate a range of light conditions, but excessive shade can reduce yields.
- **Planting** should be done at the beginning of the rainy season, when the soil is moist but not waterlogged. This allows the seedlings to establish themselves before the onset of heavy rainfall.

In general, the ideal time for planting cassava is when the weather is warm, humid, and sunny, with adequate moisture in the soil. Cassava is a tropical crop that requires favorable weather conditions to achieve high productivity.

5.4.3. Weather Conditions Contributing to High Productivity

The ideal weather conditions for high cassava productivity include:

- Warm temperatures, with average temperatures ranging from 25°C to 30°C. Temperatures above 30°C can lead to moisture stress, while temperatures below 20°C can slow down growth.
- Adequate and well-distributed rainfall, with annual totals of at least 750 mm. Consistent moisture levels, especially during the first six months after planting, are crucial for optimal growth.
- High humidity, with an average relative humidity of 60-80%. This helps maintain soil moisture and promotes healthy growth.
- Moderate solar radiation, with average daily sunshine hours of 6-8 hours. Excessive shade can reduce yields, while excessive sunlight can lead to moisture stress.
- Low wind speeds, as high winds can damage plants, disrupt soil moisture, and reduce yields.
- A dry spell of about 30-60 days before harvesting, which helps to harden the tubers and make them easier to harvest.

Cassava requires a warm, humid, and sunny climate with adequate rainfall and moderate solar radiation to achieve high productivity.

5.5. Taro (*Colocasia esculenta*)

Taro, also known as colocasia, is a tropical root crop that is widely cultivated in India (Fig. 5.6).

Fig. 5.6: Taro crop

An overview of its area, production, and importance in the country is given here under:

- **Area:** Taro is grown in various parts of India, with major production areas in the states of West Bengal, Odisha, Bihar, Assam, and Kerala. According to the Indian Council of Agricultural Research, taro is cultivated in around 0.45 million hectares of land in the country.
- **Production**: India is one of the largest producers of taro in the world. The country produces around 1.5 million tonnes of taro annually, with an average yield of around 3.3 tonnes per hectare.
- **Importance**: Taro is an important crop in India, particularly in the northeastern and eastern regions, where it is a staple food. The crop is rich in nutrients, including carbohydrates, fiber, and minerals, making it a valuable source of nutrition for millions of people. Taro is also used in various traditional dishes and is an important ingredient in many Indian cuisines. Additionally, taro is used in the production of starch, flour, and other industrial products, providing a source of income for many farmers and rural communities.

5.5.1. Agroclimatic conditions required for Taro cultivation

Taro is a tropical crop that requires specific agroclimatic conditions to grow well in India. Here are the conditions required for its cultivation:

- Taro requires a warm and humid climate with average temperatures ranging from 20°C to 30°C. The ideal temperature for taro cultivation is between 25°C and 28°C.

- The crop requires adequate rainfall, with an annual rainfall of at least 1,000 mm. However, it is sensitive to waterlogging, so well-drained soil is essential.
- Taro prefers a humid environment, with relative humidity ranging from 60% to 80%. High humidity helps to promote healthy growth and reduces the risk of pests and diseases.
- In terms of soil, taro prefers well-drained, fertile soils with a pH range of 5.5 to 6.5. It can grow in a variety of soils, including clay, loam, and sand, but the soil should be rich in organic matter.

Taro is typically planted in low-lying areas or in fields with supplemental irrigation. In areas with limited rainfall, irrigation is necessary to ensure adequate moisture levels. In India, the ideal time for planting taro is during the monsoon season, which typically begins in June or July. This allows the crop to take advantage of the abundant rainfall and warm temperatures during the growing season.

5.5.2. Optimum Weather Conditions Required fir Planting Taro

The optimum weather conditions suitable for planting taro are:

- Taro requires a warm and humid climate, with temperatures ranging from 20°C to 30°C for optimal growth. Temperatures above 35°C can lead to scorching of leaves and reduced yields.
- The ideal time for planting taro is during the monsoon season or when the soil is moist. Adequate rainfall or supplemental irrigation is necessary to keep the soil consistently moist during the first few weeks after planting.
- Relative humidity of 60-80% is ideal for taro cultivation. High humidity helps to promote healthy growth and reduces the risk of pests and diseases.
- Taro is typically planted in areas with low-lying fog or mist, which helps to maintain humidity and reduce transpiration.
- Planting taro during the monsoon season also helps to reduce soil temperature, which can become excessively hot during the summer months.

In general, taro prefers weather conditions with mild temperatures, high humidity, and adequate rainfall or irrigation.

5.5.3. Effect of Extreme Weather Conditions

Extreme weather conditions can significantly impact taro crops, leading to reduced yields, lower quality tubers, and even complete crop failure. Here are some effects of extreme weather conditions on taro crops:

- Drought can lead to reduced tuber formation and size, wilting and yellowing of leaves, and increased susceptibility to pests and diseases. Taro requires consistent moisture, especially during the first few weeks after planting.
- Flooding, on the other hand, can cause rotting of tubers and roots, reduced oxygen availability leading to root damage, and increased risk of diseases such as phytophthora and pythium. Taro is sensitive to waterlogging, making it essential to ensure well-drained soil.
- High temperatures above 35°C can cause scorching of leaves and reduced photosynthesis, reduced tuber formation and quality, and increased water loss and drought stress.
- Low temperatures below 15°C can cause reduced growth and tuber formation, increased susceptibility to diseases, and damage to leaves and stems.
- Strong winds can cause mechanical damage to leaves and stems, reduced photosynthesis and growth, and increased risk of pests and diseases.
- Hail can cause physical damage to leaves and stems, reduced photosynthesis and growth, and increased risk of pests and diseases.

Extreme weather conditions can significantly impact taro crops, and farmers should take measures to mitigate these effects, such as using irrigation, mulching, and windbreaks.

5.6. Carrots (*Daucus carota*)

Carrots are a significant crop in India (Fig. 5.7), with an estimated acreage of around 80000 hectares. This highlights their economic importance in the country.

Fig. 5.7: Carrot crop

- In terms of production, India is among the top carrot-producing countries globally. Carrots are widely cultivated in Europe, America, Asia and Africa.
- The importance of carrots in India can be attributed to several factors:
  - **Nutritional Value**: Carrots are rich in high carotene content (pro-vitamin A), which has anti-carcinogenic effects due to carotene.
  - **Economic Significance**: Carrots are among the top-ten most economically important vegetable crops in the world, both in terms of acreage and market value.
  - **Cultivation and Trade**: India's favorable climate and soil conditions make it an ideal place for carrot cultivation, with the crop being traded both domestically and internationally.

5.6.1. Agroclimatic Conditions Required to Grow Carrots

Carrots are a cool-season crop that requires specific agroclimatic conditions to grow well in India. Here are the conditions required:

- Carrots thrive in temperate climates with moderate temperatures. The ideal temperature for carrot growth is between 15°C and 20°C. Temperatures above 25°C can lead to bolting, where the plant prematurely goes to seed.
- Carrots require a relatively low humidity environment, ideally below 60%. High humidity can lead to fungal diseases.
- Carrots need adequate moisture, especially during the germination and root development stages. The ideal annual rainfall for carrot cultivation is around 600-800 mm. However, supplemental irrigation may be necessary during dry spells.
- Carrots prefer well-draining, loose soils with a pH between 6.0 and 6.8. The soil should be rich in organic matter to support healthy root development.
- In India, the ideal time to grow carrots is during the winter months (October to February), when the temperatures are cooler. In the hills, carrots can be grown during the summer months (May to July) as well.

Overall, carrots require a cool, dry, and well-draining environment to grow well in India.

5.6.2. Optimum Weather Conditions Required for Sowing

The ideal weather conditions for sowing or planting carrots in India are:

- Carrots are typically sown in the winter months, from September to November, or in early spring, from February to April, when the weather is cooler. The ideal temperature for sowing carrots is between 10°C and 20°C.
- The soil temperature should be around 15°C to 18°C for optimal germination.
- Adequate moisture is essential for germination, but excessive rainfall can lead to poor germination or seed rot. Therefore, it's crucial to sow carrots when the soil is moist but not waterlogged.
- In the hills, carrots can be sown in the summer months, from May to July, when the temperature is around 15°C to 20°C.

The ideal weather conditions for sowing carrots in India are characterized by cool temperatures, adequate moisture, and well-draining soil.

5.6.3. Weather Conditions Favorable for High Oroductivity

For high productivity of carrots in India, the following favorable weather conditions are required:

- **Cool temperatures**: Carrots thrive in temperatures ranging from 15°C to 20°C. This temperature range promotes healthy growth, root development, and sugar accumulation.
- **Moderate rainfall**: Carrots require consistent moisture, especially during the germination and root development stages. An annual rainfall of around 600-800 mm is ideal, with supplemental irrigation during dry spells.
- **Low humidity:** Carrots prefer relatively low humidity, below 60%. High humidity can lead to fungal diseases, reducing yields and affecting quality.
- **Well-distributed rainfall**: Carrots benefit from well-distributed rainfall throughout the growing season. This ensures consistent moisture levels, promoting healthy growth and development.
- **Absence of extreme weather**: Carrots are sensitive to extreme weather conditions like heavy rainfall, strong winds, and frost. Avoiding such conditions helps minimize damage and ensures high productivity.

A combination of cool temperatures, moderate rainfall, low humidity, and well-distributed rainfall, along with the absence of extreme weather conditions, is essential for achieving high carrot productivity in India.

5.6.4. Effect of Extreme Weather Events

Extreme weather events can significantly impact carrot crops, leading to reduced yields, lower quality, and even complete crop failure. Here are the effects of extreme weather events on carrot crops:

- Heavy rainfall and flooding cause seed rot and poor germination. This leads to root rot and other fungal diseases, resulting in nutrient deficiencies and reduced plant growth. Additionally, carrots may become misshapen or forked.
- Drought and water stress reduce plant growth and root development. This causes carrots to become deformed or forked and increases susceptibility to pests and diseases. As a result, yields are lower, and quality is reduced.
- High temperatures cause bolting, where the plant prematurely goes to seed. This leads to reduced sugar content and lower quality. High temperatures also increase water loss and stress, causing sun scalding and other heat-related damage.
- Frost and freezing temperatures damage or kill seedlings, resulting in blackening or browning of leaves. This can cause roots to become deformed or rot, leading to reduced yields and lower quality.
- Strong winds cause physical damage to plants, leading to soil erosion and moisture loss. This increases susceptibility to pests and diseases, resulting in reduced yields and lower quality.
- Hailstorms cause physical damage to plants and roots, leading to reduced yields and lower quality. This increases susceptibility to pests and diseases, resulting in costly repairs and replanting.

Overall, extreme weather events can have devastating effects on carrot crops, highlighting the importance of implementing climate-resilient agricultural practices and developing strategies to mitigate these impacts.

5.7. Beet (*Beta vulgaris*)

Beetroot, also known as beeds, is a root vegetable that is widely cultivated and consumed in India (Fig. 5.8). Here are some key statistics and facts about beetroot in India:

Fig. 5.8: Beet crop

- **Area and Production**: New Delhi, the capital of India, accounts for a significant portion of the country's beetroot production. According to the Indian government's agricultural data, beetroot is cultivated in approximately 40,000 hectares of land across the country. India produces around 350,000 metric tons of beetroot annually.
- **Importance**: Beetroot is a nutrient-rich vegetable, boasting high levels of vitamins, minerals, and antioxidants, making it a nutritious addition to the Indian diet. Beetroot is used in a variety of Indian dishes, such as salads, juices, pickca, and curries. For many Indian farmers, beetroot is a significant crop, providing a source of income and employment. In addition, beetroot has been used in traditional Indian medicine for its purported health benefits, including reducing blood pressure and inflammation.
- **Major Beetroot-Producing States:** Maharashtra accounts for the largest share of India's beetroot production. Uttar Pradesh is the second-largest producer of beetroot in India. Punjab is also a significant producer of beetroot in India.

Overall, beetroot is an important crop in India, providing nutritional, economic, and medicinal benefits to the country's population.

5.7.1. Agroclimatic Conditions Suitable for Growing Beets

Beets are a temperate crop that can thrive in a variety of agroclimatic conditions. Here are the suitable conditions for growing beets:

- **Temperature:** Beets require a cool and moderate temperature to grow. The ideal temperature for beet cultivation is between 10°C and 25°C. Temperatures above 30°C can lead to bolting, while temperatures below 5°C can cause damage to the crop.
- **Soil:** Beets prefer well-draining, loose, and fertile soil with a pH range of 6.0 to 7.0. The soil should be rich in organic matter to support healthy root development.

- **Moisture**: Beets require consistent moisture, especially during the germination and root development stages. The ideal annual rainfall for beet cultivation is around 500-600 mm. Supplemental irrigation may be necessary during periods of drought.
- **Sunlight**: Beets require full sun to partial shade. They should receive at least 6 hours of direct sunlight per day.
- **Region**: Beets can be grown in a variety of regions, including the plains, hills, and valleys. They are typically grown as a winter crop in the plains and as a summer crop in the hills.
- **Sowing Time**: The ideal sowing time for beets varies depending on the region. In general, beets are sown in the winter months (October-December) in the plains and in the summer months (April-June) in the hills.

By understanding these agroclimatic conditions, farmers and gardeners can create an optimal environment for growing beets.

5.7.2. Optimum Weather Conditions Required for Sowing Beet

- **Temperature**: The optimum temperature for germination of beets is between 10°C to 25°C. For growth, the ideal temperature range is 15°C to 20°C.
- **Soil**: Beets prefer well-draining, loose, and fertile soil with a pH range of 6.0 to 7.0. The soil should be rich in organic matter to support healthy root development.
- **Moisture**: Consistent moisture is essential during the germination and root development stages of beets. The ideal annual rainfall for beet cultivation is around 500-600 mm. Supplemental irrigation may be necessary during periods of drought.
- **Sunlight:** Beets require full sun to partial shade, with at least 6 hours of direct sunlight per day.
- **Region and Sowing Time:** In the plains, beets are typically sown as a winter crop between October and December. In the hills, they are sown as a summer crop between April and June. In coastal regions, the ideal sowing time is between September and November or January and February.
- **Sowing Depth**: For optimal growth, seeds should be sown 2-3 cm deep and 2-3 cm apart.

By considering these agroclimatic conditions, farmers and gardeners can optimize the sowing of beets for maximum growth and yield.

5.7.3. Effect of Extreme Weather Events

Extreme weather events can significantly impact the cultivation of beets, affecting their growth, yield, and quality. Here are some of the effects of extreme weather events on beet cultivation:

- **Drought**: Drought can lead to reduced germination and seedling establishment. This can result in stunted growth and reduced yields. Drought can also increase the susceptibility of beets to pests and diseases, ultimately reducing the sugar content and quality of the crop.
- **Floods**: Floods can result in waterlogged soil, leading to root rot and reduced yields. The increased moisture can also lead to an increased risk of disease transmission. Additionally, floods can cause nutrient deficiencies due to soil erosion, further impacting beet cultivation.
- **Heatwave:** Heatwaves can cause bolting, leading to premature flowering and reduced yields. The high temperatures can also increase transpiration, resulting in water stress. This can reduce the sugar content and quality of the beets, making them more susceptible to pests and diseases.
- **Unseasonal Rainfall**: Unseasonal rainfall can disrupt sowing and harvesting schedules, reducing seed germination and seedling establishment. The increased moisture can also lead to an increased risk of disease transmission. Furthermore, unseasonal rainfall can cause nutrient deficiencies due to soil erosion.
- **Hailstorms:** Hailstorms can result in physical damage to plants, leading to reduced yields. The damage can also increase the risk of disease transmission. Hailstorms can reduce the sugar content and quality of the beets, making them less desirable for consumption.

To mitigate the effects of extreme weather events, farmers can adopt strategies such as diversifying their crop portfolios and implementing conservation agriculture practices. Using climate-resilient crop varieties and investing in irrigation systems and water harvesting infrastructure can also help. Developing early warning systems and emergency preparedness plans can provide farmers with the necessary tools to respond to extreme weather events.

5.8. Radish (*Raphanus sativus*)

Radish is a significant root vegetable crop in India, widely cultivated and consumed across the country (Fig. 5.9). Here are some key facts about radish in India:

- **Area and Production**: India is one of the largest producers of radish in the world. According to the Indian Council of Agricultural Research

(ICAR), radish is grown in an area of around 70,000 hectares in India. The major radish-producing states in India are Maharashtra, Gujarat, Uttar Pradesh, Bihar, and West Bengal. India produces around 1.5-2 million tons of radish annually.

Fig. 5.9: Radish crop

- **Importance of Radish in India**: Radish is a popular ingredient in Indian cuisine, used in salads, curries, and as a garnish. It is rich in vitamins, minerals, and antioxidants, making it a healthy addition to the diet. Radish is an important cash crop for many farmers in India, providing a source of income. It has been used in traditional Indian medicine (Ayurveda) for its medicinal properties, including its ability to aid digestion and relieve respiratory issues. Radish is considered a sacred vegetable in some Indian cultures and is used in various rituals and ceremonies.
- **Challenges and Opportunities:** Radish cultivation is vulnerable to climate change, with rising temperatures and changing precipitation patterns affecting yields. The crop is susceptible to pests and diseases, such as aphids and powdery mildew, which can impact yields and quality. Radish prices can fluctuate significantly, affecting the livelihoods of farmers. However, India has the potential to export radish to other countries, providing a new market opportunity for farmers.

Overall, radish is an important crop in India, providing nutritional, economic, and cultural benefits. However, the crop faces challenges such as climate change, pests, and diseases, which need to be addressed to ensure sustainable production and livelihoods for farmers.

5.8.1. Agroclimatic Conditions Required for Growing Radish

Radish is a cool-season crop that requires specific agroclimatic conditions to grow well. Here are the ideal conditions for growing radishes:

- **Temperature**: Radish grows best in temperatures ranging from 15°C to 25°C. It can tolerate a wide range of temperatures, but extreme temperatures can affect its growth and quality. Temperatures above 30°C can lead to bolting, while temperatures below 10°C can slow down growth.
- **Moisture**: Radish requires adequate moisture, especially during the germination and root development stages. It thrives in well-drained soils with a consistent moisture supply. However, excessive moisture can lead to rot and other diseases.
- **Soil**: Radish can grow in a wide range of soils, but it prefers well-drained, loose soils with a pH between 6.0 and 7.0. It can tolerate slightly acidic to neutral soils, but extreme pH levels can affect its growth.
- **Sunlight:** Radish requires full sun to partial shade. It grows best in areas receiving direct sunlight for at least 4-6 hours a day.
- **Altitude:** Radish can be grown at various altitudes from sea level to 1,500 meters above sea level. However, it grows best at lower elevations, where the temperature and moisture conditions are more favorable.
- **Growing Season**: Radish is typically grown during the cooler months of the year, from September to February in the Northern Hemisphere, and from March to July in the Southern Hemisphere. It can be grown year-round in regions with mild winters and cool summers.

Radish is a versatile crop that can be grown in various agroclimatic conditions. However, it grows best in cool, moist environments with well-drained soils and full sun to partial shade.

5.8.2. Optimum Weather Conditions Required for Sowing

The optimum weather conditions for sowing radishes are:

- **Temperature**: The ideal temperature for sowing radishes is between 10°C and 20°C. Temperatures above 25°C can lead to poor germination, while temperatures below 5°C can slow down germination.
- **Soil Temperature**: The ideal soil temperature for sowing radishes is between 10°C and 15°C. Soil temperature above 20°C can lead to poor germination.
- **Moisture**: Radishes require consistent moisture during the germination period. The soil should be kept moist but not waterlogged. A relative humidity of 60-80% is ideal.
- **Weather Conditions:** Radishes prefer calm and settled weather conditions during

- In the plains: September to November and January to February
- In the hills: April to June and September to November

- **Weather Forecast:** Before sowing radishes, check the weather forecast for the next 7-10 days. Avoid sowing during periods of heavy rainfall, extreme temperatures, or strong winds.

By sowing radishes during the optimum weather conditions, you can ensure good germination, healthy growth, and a high yield.

5.8.3. Favorable Weather Conditions for High Productivity

The favorable weather conditions for high productivity of radish are:

- **Temperature**: A temperature range of 15°C to 20°C is ideal for radish growth. Temperatures between 10°C and 25°C can also support growth, but yields may be lower.
- **Moisture**: Consistent and adequate moisture is essential for radish growth. A relative humidity of 60-80% and an annual rainfall of 600-800 mm are ideal.
- **Sunshine**: Radishes require full sun to partial shade. Direct sunlight for 4-6 hours a day promotes healthy growth and development.
- **Soil Temperature**: Soil temperatures between 10°C and 20°C support optimal radish growth.
- **Weather Patterns**: Radish prefers calm and settled weather conditions. Avoid extreme weather conditions like strong winds, heavy rainfall, and extreme temperatures.
- **Growing Season**: In India, the ideal growing seasons for radishes are in the plains from September to November and January to February, and in the hills from April to June and September to November.
- **Weather Conditions During Critical Growth Stages**: During the germination stage, which lasts around 7-10 days, consistent moisture and temperatures between 10°C and 20°C are essential.
- During the root development stage, which lasts around 20-25 days, adequate moisture and temperatures between 15°C and 25°C are necessary.
- Finally, during the maturation stage, which lasts around 30-40 days, reduced moisture and temperatures between 10°C and 20°C are ideal.

By providing radishes with these favorable weather conditions, you can promote healthy growth, maximize yields, and achieve high productivity.

5.8.4. Effect of Extreme Weather Events

Extreme weather events can significantly impact radish crops, affecting their growth, development, and yield. Here are some effects of extreme weather events on radish crops:

- **Drought**: Drought can lead to reduced radish yields, poor quality, and increased susceptibility to pests and diseases. Radish requires consistent moisture, especially during the germination and root development stages. Prolonged drought can cause radish plants to become stressed, leading to a decrease in yields.
- **Floods:** Flooding can be detrimental to radish crops, causing root rot, reduced growth, and increased disease susceptibility. Excess moisture can lead to the development of fungal diseases, such as root rot and leaf spot. Additionally, flooding can cause radish plants to become waterlogged, leading to reduced yields and poor quality.
- **Heatwaves**: Heatwaves can cause radish plants to bolt, leading to reduced yields and poor quality. High temperatures can also increase the susceptibility of radish plants to pests and diseases. Radishes prefer cooler temperatures, and prolonged exposure to high temperatures can cause stress, leading to reduced growth and yields.
- **Cold Snaps**: Cold snaps can damage or kill radish plants, especially if they occur during the germination or seedling stages. Radishes are sensitive to frost, and prolonged exposure to cold temperatures can cause damage to the roots and leaves. Cold snaps can also delay radish growth, **leading to reduced yields and poor quality.**
- **Windstorms**: Windstorms can cause physical damage to radish plants, leading to reduced yields and poor quality. Strong winds can also increase the susceptibility of radish plants to pests and diseases. Additionally, windstorms can cause soil erosion, leading to reduced soil fertility and affecting radish growth.
- **Hailstorms**: Hailstorms can cause significant damage to radish plants, leading to reduced yields and poor quality. Hail can damage the leaves and roots of radish plants, making them more susceptible to pests and diseases. Hailstorms can also cause soil compaction, leading to reduced soil fertility and affecting radish growth.

The extreme weather events can have a significant impact on radish crops, affecting their growth, development, and yield. It is essential for farmers to take measures to mitigate the effects of extreme weather events, such as using irrigation systems, mulching, and crop insurance.

5.9. Turnip (*Brassica rapa*)

Turnips are a cool-season crop in India, primarily cultivated in the northern regions, particularly in the states of Punjab, Haryana, and Uttar Pradesh (Fig. 5.10).

Fig. 5.10: Turnip crop

- In terms of area, turnips are grown on approximately 80,000 hectares of land in India.
- The annual production of turnips in India is around 500,000 metric tons.
- Turnips hold importance in Indian agriculture due to their adaptability to varied climates and their role as a cash crop for many small-scale farmers. They are also a rich source of nutrients, including vitamin C, potassium, and fiber, making them a valuable addition to Indian cuisine, particularly during the winter months. turnips

5.9.1: Agroclimatic Conditions Required for Turnips

Turnips require specific agroclimatic conditions to grow well. Here are the ideal conditions:

- Turnips thrive in temperate climates with moderate temperatures. The ideal temperature for turnip growth is between 10°C and 20°C.
- Turnips require adequate moisture, especially during the germination and root development stages. Well-distributed rainfall of around 500-750 mm is ideal.
- Turnips prefer full sun to partial shade. Direct sunlight for 6-8 hours a day promotes healthy growth.
- Turnips can grow in a wide range of soils, but well-drained, loose soils with a pH between 6.0 and 7.0 are ideal.

5.9.2. Optimum Weather Conditions Required for Sowing Turnip

The optimum agroclimetic conditions for sowing turnips include:

- Temperature ranging from 10°C to 20°C, which allows for ideal germination and growth.
- Moisture levels with well-distributed rainfall of around 500-750 mm, ensuring adequate water supply during critical stages.
- Full sun to partial shade, with direct sunlight for 6-8 hours a day promoting healthy growth.
- Soil with a pH between 6.0 and 7.0, and well-drained, loose texture allowing for proper root development.
- Sowing typically takes place in late summer or early fall, about 8 to 10 weeks before the first frost, allowing the crop to mature before the onset of winter.

5.9.3. Weather Conditions Favorable for High Productivity of Turnips

For high productivity of turnips, the following weather conditions are favorable:

- **Cool temperatures**: Turnips thrive in temperatures ranging from 10°C to 20°C. Cooler temperatures promote healthy growth and prevent bolting.
- **Moderate rainfall**: Adequate and well-distributed rainfall of around 500-750 mm is essential for turnip growth. Consistent moisture levels prevent water stress and promote root development.
- **Low humidity**: Turnips are susceptible to fungal diseases, which thrive in high humidity. Low humidity helps prevent disease outbreaks.
- **Adequate sunlight**: Turnips require full sun to partial shade. Direct sunlight for 6-8 hours a day promotes healthy growth and development.
- **Frost-free conditions**: Turnips are sensitive to extreme temperatures. A frost-free period of at least 60-90 days after sowing allows the crop to mature without damage.
- **Gradual temperature decrease**: A gradual decrease in temperature during the maturation stage helps to improve the quality and sweetness of the turnips.

5.9.4. Effect of Extreme Weather Events

Extreme weather events can significantly impact turnip crops, leading to reduced yields, lower quality, and even complete crop failure. Here are some effects of extreme weather events on turnip crops:

- **Drought:** conditions lead to reduced water availability, causing stunted growth, bolting, and reduced root size. This also increases the susceptibility of the crop to pests and diseases.
- **Floods:** cause waterlogged soils, resulting in root rot and plant death. Additionally, floods lead to nutrient deficiencies and reduced microbial activity in the soil.
- **Heatwaves:** characterized by high temperatures above 25°C, cause bolting, reducing root quality and yield. Heatwaves also lead to increased moisture loss, resulting in drought stress.
- **Frost damage:** or destruction of the crop, especially during the germination and seedling stages, is a significant concern. Frost reduces growth rates and increases the susceptibility of the crop to disease.
- **Hailstorms:** cause physical damage to the crop, leading to reduced yields and lower quality. Hailstorms also increase the risk of disease infection through wounds caused by hail.
- **Strong winds:** cause physical damage to the crop, reducing yields and quality. Strong winds also increase the risk of soil erosion, especially in fields with loose or exposed soil.

To mitigate these effects, farmers can adopt strategies such as using drought-tolerant or resistant varieties, implementing conservation agriculture practices to reduce soil erosion, providing windbreaks or row covers to protect against strong winds and hail, irrigating efficiently to minimize water stress during droughts, and planting at optimal times to avoid extreme weather events.

5.10. Onions (*Allium cepa*)

India is the second-largest producer of onions in the world, accounting for over 19% of global production. Onions are a crucial crop in India, playing a significant role in the country's food security, economy, and culture (Fig. 5.11).

- **Area Under Onion Cultivation:** According to the Ministry of Agriculture and Farmers Welfare, Government of India, the area under onion cultivation in India is approximately 1.4 million hectares.
- **Production of Onion in India**: India produces over 22 million tons of onions every year, with the majority coming from the states of Maharashtra, Madhya Pradesh, Karnataka, Gujarat, and Uttar Pradesh.
- **Importance of Onion in India**: Onions are a staple ingredient in Indian cuisine, and their availability is crucial for food security. Onion production provides livelihood support to millions of farmers, laborers, and traders

across the country. India exports onions to several countries, including Bangladesh, Malaysia, and the Middle East, earning significant foreign exchange. Onions also play a vital role in Indian culture, particularly in Hindu and Muslim traditions, where they are used in various rituals and ceremonies.

Fig. 5.11: Onion crop

- **Challenges Faced by Onion Farmers in India**: Onion prices in India are highly volatile, affecting farmers' incomes and livelihoods. Climate change has impacted onion production, leading to reduced yields and lower quality onions. Onions are also susceptible to various diseases and pests, which can significantly reduce yields and affect quality.
- **Government Initiatives to Support Onion Farmers**: The government announces a Minimum Support Price (MSP) for onions to ensure farmers receive a fair price for their produce. The government also provides crop insurance to onion farmers to protect them against crop failures and price fluctuations. Additionally, the government supports research and development initiatives to improve onion productivity, disease resistance, and quality.

Onions play a vital role in India's food security, economy, and culture. The government's initiatives and support are crucial in addressing the challenges faced by onion farmers and ensuring the sustainability of onion production in India. Onions are a cool-season crop, and their growth is influenced by various agroclimatic factors.

5.10.1. Agroclimatic Conditions Required for Growing Onions

The ideal conditions for growing onions in India are:

- **Temperature**: The optimum temperature for onion growth is between 13°C to 24°C. Seed germination occurs best between 15°C to 20°C, while bulb formation takes place between 15°C to 18°C.

- **Moisture**: Onions require adequate moisture, especially during the germination and bulb formation stages. The ideal rainfall for onion cultivation is between 500 mm to 750 mm per annum. Consistent moisture, especially during the first 30 days after sowing, is crucial for onion growth.
- **Soil:** Well-draining, fertile soils with a pH range of 6.0 to 7.0 are ideal for onion cultivation. Soils with high organic matter content and good water-holding capacity are also suitable.
- **Sunlight:** Onions require full sun to partial shade. The ideal duration of sunlight for onion growth is between 10 to 12 hours per day.
- **Region-wise Agroclimatic Conditions:** India's diverse agroclimatic regions support onion cultivation. The main onion-growing regions are Maharashtra, Madhya Pradesh, Karnataka, and Gujarat. Maharashtra is characterized by a hot and dry climate, with temperatures ranging from 15°C to 35°C. Madhya Pradesh has a subtropical climate with temperatures ranging from 10°C to 40°C. Karnataka experiences a tropical climate with temperatures ranging from 15°C to 30°C. Gujarat has a hot and dry climate, with temperatures ranging from 15°C to 40°C.
- **Season-wise Agroclimatic Conditions**: Onion cultivation in India is done during two main seasons: Rabi (October to March) and Kharif (June to September). The Rabi season is characterized by cool temperatures and low humidity, ideal for onion growth. The Kharif season is marked by high temperatures and humidity, suitable for certain onion varieties.

Understanding the agroclimatic conditions required for onion cultivation is crucial for optimizing crop growth, yield, and quality in India's diverse regions and seasons.

5.10.2. Optimum Weather Conditions Required for Sowing Onions

Onions are a cool-season crop, and their growth is influenced by various agroclimatic factors. The ideal conditions for sowing onions in India are:

- **Temperature**: The optimum temperature for sowing onions is between 15°C to 20°C. Temperatures above 25°C can lead to poor germination and reduced yields.
- **Moisture**: Adequate moisture is essential for germination and seedling establishment. The ideal rainfall for sowing onions is between 500 mm to 750 mm per annum. Consistent moisture, especially during the first 30 days after sowing, is crucial for onion growth.

- **Soil:** Well-draining, fertile soils with a pH range of 6.0 to 7.0 are ideal for onion cultivation. Soils with high organic matter content and good water-holding capacity are also suitable.
- **Sunlight:** Onions require full sun to partial shade. The ideal duration of sunlight for onion growth is between 10 to 12 hours per day.
- **Sowing Time:** The ideal time for sowing onions in India varies depending on the region. In general, onions are sown during the Rabi season (October to March) in the northern and western regions, and during the Kharif season (June to September) in the southern and eastern regions.
- **Region-wise Sowing Time**: Maharashtra and Gujarat: September to October
  - Madhya Pradesh and Uttar Pradesh: October to Novembery
  - Karnataka and Tamil Nadu: August to September
  - Bihar and West Bengal: October to November

Understanding the agroclimatic conditions required for sowing onions is crucial for optimizing crop growth, yield, and quality in India's diverse regions.

5.10.3. Favorable Weather Conditions for High Productivity

Onions are a cool-season crop, and their growth is influenced by various weather factors. The following favorable weather conditions can lead to high onion productivity:

- **Temperature:** Onions grow best in temperatures ranging from 13°C to 24°C. A temperature range of 15°C to 18°C is ideal for bulb formation. Avoid exposing onions to extreme temperatures above 30°C or below 10°C.
- **Moisture:** Adequate and consistent moisture is essential for onion growth. Onions require about 500 mm to 750 mm of rainfall per annum. Irrigation can supplement rainfall, but avoid overwatering, which can lead to rot and other diseases.
- **Sunshine**: Onions require full sun to partial shade. The ideal duration of sunlight for onion growth is between 10 to 12 hours per day.
- **Wind**: Onions prefer a moderate wind speed of about 10 km/h. High winds can damage the plants, while low winds can lead to fungal diseases.
- **Relative Humidity** : A relative humidity of 60% to 80% is ideal for onion growth. High humidity can lead to fungal diseases, while low humidity can cause moisture stress.

- **Weather Conditions During Growth Stages**
  - Germination stage: Cool temperatures (15°C to 20°C) and consistent moisture.
  - Seedling stage: Moderate temperatures (15°C to 25°C) and adequate moisture.
  - Bulb formation stage: Cooler temperatures (15°C to 18°C) and reduced moisture.
  - Maturation stage: Dry and sunny weather to promote drying and curing of the bulbs.

By providing onions with these favorable weather conditions, farmers can promote healthy growth, maximize yields, and achieve high productivity.

5.10.4. Effect of Extreme Weather Events

Onion crops are highly susceptible to extreme weather events, which can significantly impact yield, quality, and market prices. The following extreme weather events can affect onion crops:

- **Drought:** Drought can lead to reduced onion yields, smaller bulb sizes, and lower quality. Onions require consistent moisture, especially during the germination and bulb formation stages. Prolonged drought can cause onions to become stressed, making them more vulnerable to pests and diseases.
- **Floods**: Excessive rainfall and flooding can cause onions to rot, leading to significant yield losses. Onions are sensitive to waterlogging, which can cause the bulbs to become soft and rot. Flooding can also lead to the spread of diseases, such as fungal infections.
- **Heatwaves**: High temperatures during heatwaves can cause onions to bolt, or go to seed, reducing yields and quality. Onions are sensitive to temperatures above 30°C, which can cause the plants to become stressed and lead to reduced growth.
- **Unseasonal Rainfall**: Unseasonal rainfall during the harvest season can cause onions to rot, leading to significant yield losses. Onions are sensitive to moisture during the harvest season, and excessive rainfall can cause the bulbs to become soft and rot.
- **Hailstorms**: Hailstorms can cause significant damage to onion crops, leading to reduced yields and lower quality. Hail can damage the leaves and bulbs, making them more vulnerable to pests and diseases.
- **Cyclones**: Cyclones can cause significant damage to onion crops, leading to reduced yields and lower quality. Strong winds and excessive rainfall can cause onions to become uprooted, leading to significant yield losses.

To mitigate the effects of extreme weather events on onion crops, farmers can use various strategies, such as crop insurance, irrigation management, and crop diversification. Additionally, implementing climate-resilient agricultural practices can help reduce the vulnerability of onion crops to extreme weather events.

5.11. Garlic (*Allium sativum*)

India is a significant producer of garlic, with major garlic-producing states including Maharashtra, Gujarat, Madhya Pradesh, Rajasthan, and Uttar Pradesh (Fig. 5.12.).

- In terms of area, garlic is cultivated across approximately 200,000 hectares of land in India. The production of garlic in India stands at around 1.4 million metric tons annually.
- Garlic holds great importance in Indian cuisine and culture. It is a fundamental ingredient in many traditional Indian recipes, adding flavor and aroma to various dishes. Additionally, garlic is known for its medicinal properties and is used in Ayurvedic medicine. The demand for garlic in India is high, driven by its culinary and therapeutic uses.

Fig. 5.12: Garlic crop

5.11.1. Agroclimatic conditions suitable for growing garlic

Garlic is typically grown in regions with a cool and dry climate. The ideal agroclimatic regions for growing garlic include:

- Temperate regions with moderate winters and warm summers
- Regions with low humidity and well-defined seasonal changes
- Areas with average temperatures ranging from 10°C to 25°C (50°F to 77°F)

- In India, the suitable agroclimatic regions for garlic cultivation are:
  - The Indo-Gangetic Plains
  - The Irrigated Plains of Maharashtra and Gujarat
  - The Plateau Regions of Madhya Pradesh and Chhattisgarh
  - The Hills of North India, including Jammu and Kashmir, Himachal Pradesh, and Uttarakhand

5.11.2. Optimum Weather Conditions Required for Sowing

The optimum weather conditions required for sowing garlic are characterized by cool temperatures, moderate humidity, and adequate moisture.

- Garlic is typically sown in the winter season, from October to November in the northern plains of India, and from September to October in the hills and plateau regions.
- The ideal temperature for sowing garlic ranges from 10°C to 20°C. The soil should be moist but not waterlogged, with a relative humidity of around 60-70%. Garlic requires a period of cold weather, known as vernalization, to form bulbs, and this typically occurs when the temperature drops below 10°C for a few weeks.

5.11.3. Weather Conditions Favorable for High Productivity

For high productivity in garlic cultivation, the following weather conditions are favorable:

- Cool winters with temperatures ranging from 10°C to 15°C, which helps in the formation of bulbs
- A moderate rise in temperature during the growing period, with average temperatures between 15°C to 25°C.
- Low humidity during the maturation phase to prevent fungal diseases
- Adequate moisture during the growing period, with an annual rainfall of around 600-800 mm or supplementary irrigation
- A dry spell before harvesting to facilitate easy lifting and drying of the bulbs.

5.11.4. Effect of Extreme Weather Events

Extreme weather events can have a significant impact on garlic cultivation.

- Prolonged periods of heavy rainfall can lead to waterlogging, causing roots to rot and reducing bulb formation. Excessive moisture can also increase the risk of fungal diseases.

- Drought, on the other hand, can result in stunted growth, reduced bulb size, and lower yields. Garlic requires consistent moisture, especially during the bulb formation stage.
- Unseasonal frosts can damage or destroy garlic crops, while extreme heat waves can accelerate maturity, reducing bulb quality and size.
- Flooding can also lead to soil erosion, nutrient depletion, and contamination, affecting soil health and future garlic crops.

5.12. Ginger (*Zingiber officinale*)

India is a major ginger-producing country, with an estimated 2.43 million tons produced in 2023, cultivated across 141,000 hectares, and Madhya Pradesh being the top-producing state (Fig. 5.13).

Fig. 5.13: Ginger crop

Ginger holds significant importance in India, both culturally and medicinally. It has been an integral part of Indian cuisine and traditional medicine, Ayurveda, for centuries.

- In Ayurveda, ginger is considered a universal medicine due to its numerous health benefits. It is believed to possess natural anti-inflammatory properties, aid digestion, and relieve nausea. Ginger is also used to treat various ailments, such as colds, coughs, and respiratory issues.
- In Indian cuisine, ginger is a fundamental ingredient, adding flavor and aroma to various dishes, including curries, stir-fries, and teas. It is also used in traditional Indian beverages like chai (spiced tea) and jal jeera (spiced cumin drink).
- Furthermore, ginger holds cultural and spiritual significance in India. In Hinduism, ginger is considered a sacred plant and is often used in religious rituals and ceremonies.

Overall, ginger plays a vital role in Indian culture, medicine, and cuisine, and its importance is deeply rooted in the country's traditions and customs.

5.12.1. Agroclimatic Conditions Required for Growing Ginger

Ginger is a tropical crop that thrives in warm and humid climates. To grow ginger, specific agroclimatic conditions are required.

- Ginger requires a warm and humid climate with average temperatures ranging from 20°C to 30°C. Temperatures above 35°C can lead to sun scorching and reduced yields. The ideal annual rainfall for ginger cultivation is between 1,500 mm to 2,000 mm, with a well-distributed rainfall pattern throughout the growing season.
- Ginger also requires high humidity, ideally above 60%, to thrive. The crop is sensitive to wind, which can cause damage to the leaves and stems. Therefore, a wind-protected area is essential for ginger cultivation.
- In terms of soil, ginger prefers well-draining, fertile soil with a pH range of 5.5 to 6.5. The soil should be rich in organic matter, and a mix of sand, silt, and clay provides the ideal texture for ginger growth.

Overall, the ideal agroclimatic conditions for growing ginger include a warm and humid climate, adequate rainfall, and well-draining fertile soil.

5.12.2. Weather Conditions Required for Sowing Ginger

The optimum weather conditions required for sowing ginger are:

- Ginger is typically sown during the spring or early summer months when the weather is warm and humid. The ideal time for sowing ginger is when the soil temperature has warmed up to at least 15°C to 20°C.
- Adequate moisture in the soil is essential for germination, and therefore, sowing is usually done just before the onset of the monsoon season or during a period of consistent rainfall.
- Cloudy weather with moderate temperatures and high humidity is ideal for sowing ginger. Direct sunlight can inhibit germination, so it's best to sow ginger in a shaded area or under a thin layer of mulch.
- It's also essential to avoid sowing ginger during periods of extreme weather conditions such as heavy rainfall, drought, or intense heat.

5.12.3. Favorable Weather Conditions Required for High Oriductivity

For high productivity in ginger cultivation, the following favorable weather conditions are essential:

- Warm temperatures, ranging from 20°C to 30°C, promote healthy growth and development. Temperatures above 35°C can lead to sun scorching, while temperatures below 15°C can slow down growth.

- Adequate and well-distributed rainfall, ranging from 1,500 mm to 2,000 mm per annum, ensures optimal growth. Ginger requires consistent moisture, especially during the rhizome formation stage.
- High humidity, above 60%, favors healthy growth and development. Low humidity can lead to moisture stress, affecting yields.
- Moderate wind speeds, below 15 km/h, prevent damage to the plants and promote healthy growth.
- Adequate sunlight, especially during the initial stages of growth, promotes healthy development. However, extreme sunlight can cause scorching, so shade management is essential.
- A frost-free period, as ginger is highly sensitive to frost, is crucial for high productivity.

5.12.4. Effect of Extreme Weather Events

Extreme weather events can have devastating effects on ginger crops, leading to significant reductions in yield and quality.

- Prolonged droughts can cause moisture stress, leading to stunted growth, reduced rhizome formation, and increased susceptibility to pests and diseases. On the other hand, excessive rainfall and flooding can cause waterlogged soil conditions, leading to root rot and other fungal diseases.
- Extreme temperatures, such as heatwaves, can cause sun scorching, reduce growth rates, and lead to a decrease in rhizome quality. Conversely, unseasonal frosts can damage or kill ginger plants, resulting in significant yield losses.
- Strong winds and storms can cause physical damage to ginger plants, breaking or uprooting them, and reducing yields. Additionally, extreme weather events can disrupt the delicate balance of ecosystems, leading to an increase in pest and disease pressure.

5.13. Conclusion

Overall, extreme weather events can have far-reaching and devastating impacts on ginger crops, highlighting the need for climate-resilient agricultural practices and effective crop management strategies. In conclusion, the agroclimatology of tuber and root vegetables plays a vital role in determining the suitability and productivity of these crops in various regions. Understanding the complex interactions between climate, soil, and crop growth is essential for optimizing crop yields, improving quality, and reducing losses. The specific agroclimatic requirements of different tuber and root vegetables, such as potatoes, sweet potatoes, cassava, yams, ginger, and turmeric, have been discussed in detail.

- It is evident that temperature, rainfall, humidity, wind, and sunlight are critical climatic factors that influence the growth and development of these crops. The optimal agroclimatic conditions for each crop vary, and any deviation from these conditions can significantly impact yields and quality. Furthermore, extreme weather events, such as droughts, floods, heatwaves, and frosts, can have devastating effects on tuber and root vegetable crops, highlighting the need for climate-resilient agricultural practices.
- To improve the productivity and sustainability of tuber and root vegetable crops, it is essential to adopt agroclimatic zoning, a technique that involves dividing a region into zones based on their agroclimatic characteristics. This approach enables farmers to select the most suitable crops for their specific region, optimize crop management practices, and reduce the risks associated with climate variability.
- In addition, the use of climate-smart agricultural practices, such as conservation agriculture, agroforestry, and irrigation management, can help to mitigate the impacts of climate change on tuber and root vegetable crops. These practices promote soil health, improve water use efficiency, and enhance crop resilience to extreme weather events.

In summary, the agroclimatology of tuber and root vegetables is a critical component of sustainable agricultural production. By understanding the complex interactions between climate, soil, and crop growth, farmers and policymakers can work together to develop and implement effective strategies for improving crop productivity, reducing losses, and promoting climate resilience in the face of a changing climate.

6

Agroclimatology of Leafy Greens

Leafy greens, comprising a diverse array of crops such as spinach, lettuce, kale, collard greens, and mustard greens, among others, are a vital component of global vegetable production. With their high nutritional value, versatility in culinary uses, and ease of cultivation, leafy greens (Fig. 6.1) have become an integral part of diets worldwide.

Fig. 6.1: Leafy green vegetables

- In India, leafy greens are grown across various agroclimatic zones, with major producing states including Maharashtra, Gujarat, Uttar Pradesh, and Tamil Nadu. The estimated area under leafy greens cultivation in India ranges from 0.7 to 0.8 million hectares, accounting for a significant share of the country's total vegetable production. The annual production of leafy greens in India stands at approximately 5-6 million tonnes, with a considerable portion being consumed domestically.

- Beyond their culinary and nutritional significance, leafy greens also hold substantial economic importance, particularly for small and marginal farmers who constitute the backbone of India's agricultural sector. The cultivation of leafy greens provides these farmers with a vital source of income, enabling them to support their livelihoods and contribute to local economic development.
- However, leafy greens are highly sensitive to environmental factors, making agroclimatology a critical component of their cultivation. Temperature fluctuations, rainfall patterns, solar radiation, and other agroclimatic elements significantly influence the growth, development, and yield of leafy greens. Understanding the intricate relationships between these crops and their environment is essential for developing effective production strategies, improving crop resilience, and ensuring sustainable leafy greens production.

This chapter on agroclimatology of leafy greens aims to provide a comprehensive overview of the complex interactions between these crops and their environment. It will delve into the specific agroclimatic requirements of various leafy greens crops, explore the impact of climate change on their production, and discuss strategies for mitigating these impacts and promoting sustainable leafy greens cultivation.

6.1. Spinach (*Spinacia oleracea*)

Spinach is an essential leafy vegetable crop in India, widely cultivated and consumed across the country (Fig. 6.2). Here's an overview of its area, production, and importance:

Fig. 6.2: Spinach crop

- **Area:** Spinach is grown in various parts of India, with major production concentrated in the states of Maharashtra, Gujarat, Uttar Pradesh, and Punjab. According to the Ministry of Agriculture and Farmers Welfare, Government of India, the total area under spinach cultivation in India is around 24,000 hectares.

- **Production**: India is one of the significant producers of spinach globally. The country's spinach production has shown a steady increase over the years, thanks to improved agricultural practices, better crop varieties, and favorable climate conditions. As per the data from the Ministry of Agriculture and Farmers Welfare, India produced around 1.3 million metric tons of spinach in 2020-21.
- **Importance:** Spinach holds immense nutritional and economic importance in India. It is an excellent source of essential vitamins, minerals, and antioxidants, making it a vital component of a healthy diet. Spinach is rich in iron, calcium, and vitamins A and K, which are crucial for maintaining good health and preventing various diseases.
- In terms of economic importance, spinach is a valuable crop for many farmers in India, providing them with a lucrative source of income. The crop is in high demand, both domestically and internationally, which has led to increased exports and better prices for farmers. Additionally, spinach is also used in various processed food products, such as frozen spinach, spinach puree, and canned spinach, which contributes to the country's food processing industry.

Overall, spinach is a vital crop in India, offering numerous health benefits, economic opportunities, and contributing to the country's food security. Spinach is a cool-season crop that thrives in temperate climates with moderate temperatures and adequate moisture

6.1.1. Agroclimatic Conditions Favorable to Grow Spinach

- **Temperature**: Spinach requires a cool temperature to grow, typically between 10°C and 25°C. The ideal temperature for spinach germination is around 10°C to 15°C, while the optimal temperature for growth is between 15°C and 20°C. Temperatures above 25°C can lead to bolting, which reduces the quality and yield of the crop.
- **Moisture:** Spinach needs consistent moisture, especially during the germination and seedling stages. The crop requires around 400-600 mm of rainfall or irrigation per growing season. However, excessive rainfall or waterlogging can lead to root rot and other diseases.
- **Soil:** Spinach prefers well-draining, fertile soils with a pH range of 6.0-7.0. The crop can grow in a wide range of soils, including clay, loam, and sandy soils, but it performs best in soils with good organic matter content.
- **Day Length**: Spinach is a short-day crop, meaning it requires shorter days to grow. The ideal day length for spinach is around 10-12 hours.

Longer days can induce bolting, which reduces the quality and yield of the crop.

- **Altitude:** Spinach can be grown at various altitudes, but it performs best at elevations between 500-2000 meters above sea level. At higher elevations, the cooler temperatures and reduced humidity create ideal conditions for spinach growth.
- **Seasonality**: In India, spinach is typically grown during the winter months (October to February), when the temperatures are cooler. In regions with mild winters, spinach can be grown year-round, except during the hottest summer months.

Overall, spinach requires a cool and moist climate with well-draining soils to grow. The crop is sensitive to extreme temperatures, excessive rainfall, and waterlogging, which can impact its growth and yield.

6.1.2. Optimum Weather Conditions for Sowing Spinach

The optimum weather conditions for sowing spinach are crucial for its germination, growth, and yield. Here's a detailed overview:

- Temperature is a critical factor for spinach germination and growth. The ideal temperature for sowing spinach is when the soil temperature ranges from 10°C to 15°C. At this temperature, spinach seeds germinate rapidly, and the seedlings develop quickly.
- In terms of air temperature, spinach prefers cooler conditions, typically ranging from 15°C to 20°C during the day and around 5°C to 10°C at night. Avoid sowing spinach when the air temperature is expected to rise above 25°C, as high temperatures can lead to bolting.
- Moisture is another essential factor for spinach germination and growth. Spinach requires consistent moisture, especially during the germination and seedling stages. The ideal time for sowing spinach is when the soil is moist but not waterlogged. A well-draining soil with adequate moisture helps prevent seed rot and promotes healthy germination.
- Rainfall and irrigation also play a crucial role in spinach growth. Spinach requires around 400-600 mm of rainfall or irrigation per growing season. However, excessive rainfall or irrigation can lead to waterlogging, which can cause root rot and other diseases.
- In terms of humidity, spinach prefers a relatively low humidity, typically ranging from 50% to 70%. High humidity can lead to fungal diseases, while low humidity can cause moisture stress.

- Wind is another factor to consider when sowing spinach. Spinach is a delicate crop and can be damaged by strong winds. Avoid sowing spinach in areas with high wind speeds, as this can lead to soil erosion, reduced germination, and poor growth.
- In India, the ideal time for sowing spinach is during the winter months (October to February), when the temperatures are cooler. In regions with mild winters, spinach can be grown year-round, except during the hottest summer months.

Overall, the optimum weather conditions for sowing spinach involve a combination of cool temperatures, consistent moisture, and moderate humidity. By sowing spinach during the right weather conditions, farmers can promote healthy germination, growth, and yield.

6.1.3. Favorable Weather Conditions for High Productivity

For spinach to achieve high productivity, it requires a combination of favorable weather conditions. Here's a detailed overview:

- Temperature plays a crucial role in spinach productivity. The ideal temperature for spinach growth is between 15°C and 20°C. Within this range, spinach plants grow rapidly, and leaf production is maximized. Temperatures above 25°C can lead to bolting, reducing leaf quality and yield.
- Moisture is another critical factor for spinach productivity. Spinach requires consistent and adequate moisture, especially during the vegetative growth stage. A well-distributed rainfall or irrigation of around 400-600 mm per growing season is ideal. However, excessive moisture can lead to waterlogging, causing root rot and reducing productivity.
- Humidity also affects spinach productivity. A relative humidity of 50-70% is ideal for spinach growth. High humidity can lead to fungal diseases, while low humidity can cause moisture stress.
- Solar radiation is essential for spinach growth, but excessive radiation can cause stress. Spinach requires bright, indirect sunlight, especially during the seedling stage. As the plants mature, they can tolerate more direct sunlight.
- Wind can also impact spinach productivity. While some wind is beneficial for pollination and disease control, strong winds can damage plants, reduce growth, and lower yields.
- In terms of weather patterns, spinach productivity is favored by:

- A cool and dry winter, which allows for healthy growth and development.
- A gradual increase in temperature during spring, promoting healthy leaf production.
- Adequate moisture during the growing season, either through rainfall or irrigation.
- Avoidance of extreme weather events, such as heavy rainfall, hail, or intense heatwaves.

By providing spinach with these favorable weather conditions, farmers can promote healthy growth, maximize leaf production, and achieve high productivity.

6.1.4. Effect of Extreme Weather Events

Extreme weather events can have a devastating impact on spinach crops, leading to significant reductions in yield and quality. Here's a detailed overview of the effects of extreme weather events on spinach:

- **Heavy Rainfall and Flooding**: Heavy rainfall and flooding can be detrimental to spinach crop. Excess moisture can cause waterlogging, leading to root rot and other diseases. The excess water can also cause the soil to become water-saturated, making it difficult for the roots to breathe. This can lead to a reduction in plant growth, yellowing of leaves, and a decrease in yield.
- **Drought and Heat Stress**: Drought and heat stress can also have a significant impact on spinach crops. Spinach requires consistent moisture to grow, and drought can cause the plants to become stressed. High temperatures can also cause the plants to bolt, leading to a reduction in leaf quality and yield. Prolonged drought and heat stress can lead to plant death, resulting in significant yield losses.
- **Hail Storms**: Hail storms can cause physical damage to spinach plants, leading to bruised and torn leaves. This can reduce the quality of the leaves and make them more susceptible to disease. In severe cases, hail storms can destroy entire spinach crops, leading to significant economic losses.
- **Intense Heatwaves:** Intense heatwaves can cause spinach plants to become stressed, leading to a reduction in growth and yield. High temperatures can also cause the plants to bolt, leading to a reduction in leaf quality. Prolonged exposure to intense heat can lead to plant death, resulting in significant yield losses.

- **Strong Winds**: Strong winds can cause physical damage to spinach plants, leading to bruised and torn leaves. This can reduce the quality of the leaves and make them more susceptible to disease. In severe cases, strong winds can destroy entire spinach crops, leading to significant economic losses.
- **Frost and Freezing Temperatures**: Frost and freezing temperatures can cause significant damage to spinach crops. Spinach is a cool-season crop and can tolerate light frosts, but prolonged exposure to freezing temperatures can cause significant damage. Frost can cause the leaves to become damaged, leading to a reduction in quality and yield.

Overall, extreme weather events can have a significant impact on spinach crops, leading to reductions in yield and quality. It's essential for farmers to take measures to mitigate the effects of extreme weather events, such as using irrigation systems, row covers, and other forms of protection.

Duration: Normally it takes forty to 50 days for harvest.

6.2. Amaranth (*Amaranthus* spp)

Amaranth is a nutrient-rich, gluten-free crop that has been cultivated for thousands of years in India (Fig. 6.3). Here's an in-depth look at its area, production, and importance in India:

Fig. 6.3: Amaranth crop

- **Area of Cultivation**: Amaranth is grown in various parts of India, with a significant presence in the states of Madhya Pradesh, Rajasthan, and Uttar Pradesh. These states account for over 70% of the country's total amaranth production.
- **Production** : According to the Ministry of Agriculture and Farmers Welfare, Government of India, the production of amaranth in India was approximately 1.4 million metric tons in 2020-21. The country's amaranth production has been steadily increasing over the years, driven by growing demand for gluten-free and nutritious food products.

- **Importance**: Amaranth is a vital crop in India for several reasons:
  - **Nutritional Value**: Amaranth is an excellent source of protein, fiber, and minerals like iron, calcium, and potassium. It is also rich in antioxidants and has been linked to several potential health benefits, including reducing inflammation and improving cardiovascular health.
  - **Gluten-Free**: Amaranth is gluten-free, making it an excellent option for people with gluten intolerance or celiac disease.
  - **Food Security**: Amaranth is a drought-tolerant crop that can thrive in challenging environmental conditions, making it an essential crop for ensuring food security, particularly in regions prone to drought.
  - **Economic Benefits**: Amaranth is a valuable crop for farmers in India, providing a source of income and employment. Many small-scale farmers in the country rely on amaranth as a primary crop.
  - **Cultural Significance:** Amaranth has cultural and spiritual significance in many Indian societies. In some parts of the country, amaranth is considered a sacred crop and is used in traditional rituals and ceremonies.

In summary, amaranth is a highly nutritious, gluten-free crop that plays a vital role in ensuring food security, providing economic benefits, and holding cultural significance in India. Its ability to thrive in challenging environmental conditions makes it an essential crop for many farmers in the country.

6.2.1. Agroclimatic Conditions Required for Growing Amatanth

Amaranth is a versatile crop that can be grown in various agroclimatic conditions in India. However, it thrives best in certain conditions. Here are the ideal agroclimatic conditions required for growing amaranth in India:

- **Climate**: Amaranth requires a warm and dry climate to grow. It is a tropical crop that thrives in temperatures ranging from 25°C to 35°C. However, it can tolerate temperatures up to 40°C.
- **Soil:** Amaranth can grow in a wide range of soils, but it prefers well-drained, fertile soils with a pH range of 6.0 to 8.0. It can tolerate saline and alkaline soils, but its growth is affected in waterlogged conditions.
- **Rainfall:** Amaranth requires adequate rainfall, but it is also drought-tolerant. It requires a minimum rainfall of 500 mm to 600 mm per annum, but it can grow well with supplemental irrigation.
- **Altitude**: Amaranth can grow at various altitudes, but it is typically grown in the plains and low-altitude areas. It can grow up to an altitude of 1,500 meters above sea level.

- **Sunlight**: Amaranth requires full sunlight to grow. It should be planted in areas that receive direct sunlight for at least 6 hours a day.
- **Sowing Time:** The ideal sowing time for amaranth in India varies depending on the region:
  - In the northern plains, amaranth is typically sown in late spring to early summer (May-June).
  - In the southern states, amaranth is sown in late summer to early autumn (September-October).
  - In the eastern states, amaranth is sown in late autumn to early winter (November-December).

Overall, amaranth is a versatile crop that can grow in various agroclimatic conditions in India. However, it requires a warm and dry climate, well-drained soil, and adequate

6.2.2. Optimum Weather Conditions for Sowing

The optimum weather conditions required for sowing amaranth in India are as follows:

- Amaranth requires a warm and dry climate to grow. The ideal temperature for sowing amaranth ranges from 25°C to 35°C. It can tolerate temperatures up to 40°C, but temperatures above 40°C can lead to flower drop and reduced seed set.
- The crop requires adequate moisture during the germination and seedling stages. A well-distributed rainfall of 500 mm to 600 mm per annum is ideal for amaranth cultivation. However, it is also drought-tolerant and can grow well with supplemental irrigation.
- Amaranth is typically sown during the late spring to early summer season in the northern plains, late summer to early autumn in the southern states, and late autumn to early winter in the eastern states.
- The ideal time for sowing amaranth is when the soil has warmed up to at least 20°C and the air temperature is around 25°C to 30°C. Sowing during this period allows the crop to take advantage of the warm weather and adequate moisture, leading to optimal growth and seed production.
- It's essential to avoid sowing amaranth during periods of heavy rainfall or when the soil.

6.2.3. Weather Conditions Favorable for High Productivity

The weather conditions favorable for high productivity of amaranth in India are as follows:

- Amaranth requires a warm and dry climate with plenty of sunshine to produce high yields. The ideal temperature for high productivity ranges from 25°C to 30°C during the day and around 15°C to 20°C during the night.
- Adequate moisture is essential for high productivity, but excessive rainfall can lead to reduced yields. A well-distributed rainfall of 500 mm to 600 mm per annum is ideal for amaranth cultivation.
- Amaranth is a short-day crop, meaning it requires a relatively short period of daylight to flower and produce seeds. The ideal photoperiod for amaranth is around 12-13 hours of daylight.
- Low humidity and moderate wind speeds are also beneficial for high productivity. Amaranth is susceptible to fungal diseases, which thrive in high-humidity environments. Moderate wind speeds help to prevent fungal diseases by promoting air circulation and drying the leaves.

Overall, the weather conditions that favor high productivity of amaranth in India include warm temperatures, adequate moisture, plenty of sunshine, and low humidity.

6.2.4. Effect of Extreme Weather Conditions

Extreme weather conditions can have a significant impact on amaranth production in India.

- **Drought**: Amaranth is relatively drought-tolerant, but prolonged drought can lead to reduced yields and lower seed quality. Drought stress can also cause the plants to become more susceptible to pests and diseases.
- **Excessive Rainfall**: Excessive rainfall can lead to waterlogging, which can cause the roots to rot and reduce yields. Amaranth is also susceptible to fungal diseases, which thrive in high-moisture environments.
- **High Temperatures:** High temperatures above 40°C can lead to flower drop and reduced seed set, resulting in lower yields. Prolonged exposure to high temperatures can also cause the plants to become stressed, making them more susceptible to pests and diseases.
- **Low Temperatures:** Low temperatures below 15°C can slow down the growth and development of amaranth plants, leading to reduced yields. Frost can also damage the plants, causing them to become brittle and break.
- **Flooding**: Flooding can be disastrous for amaranth crops, causing the plants to become waterlogged and reducing yields. Flooding can also lead to soil erosion, nutrient depletion, and increased susceptibility to pests and diseases.

Overall, extreme weather conditions can have a significant impact on amaranth production in India, leading to reduced yields, lower seed quality, and increased susceptibility to pests and diseases.

Duration: 25 to 40 days and the plants can be harvested after reaching 25 to 30 Cm height.

6.3. Fenugreek (*Trigonella foenum-graecum*)

Fenugreek leaves, also known as methi leaves, are a popular vegetable in Indian cuisine (Fig. 6.4).

Fig. 6.4: Fenugreek crop

They are rich in nutrients, fiber, and antioxidants, making them a healthy addition to various dishes.

- **Culinary Uses**: Fenugreek leaves are used fresh or dried in a variety of Indian recipes. They have a slightly bitter taste, which is often balanced by combining them with spices, herbs, and other ingredients. Some popular dishes that feature fenugreek leaves include saag paneer, a creamy spinach curry with paneer (Indian cheese), methi leaves, and spices. Methi chaman is another popular dish, which is a spicy curry made with fenugreek leaves, paneer, and spices. Fenugreek leaves are also used to make methi dal, a flavorful lentil curry made with fenugreek leaves, split red lentils, and spices. Additionally, methi roti is a type of Indian flatbread flavored with fenugreek leaves and spices.
- **Nutritional Benefits**: Fenugreek leaves are rich in vitamins A, C, and K, as well as minerals like calcium, iron, and potassium. They are also a good source of fiber and antioxidants, and contain phytochemicals that have anti-inflammatory and anti-diabetic properties.
- **Health Benefits:** Consuming fenugreek leaves has been associated with several health benefits. They have been shown to lower blood sugar levels, reduce inflammation and improve digestion, and support heart health by reducing cholesterol levels. Fenugreek leaves have also

been used as a natural remedy for respiratory issues like bronchitis and asthma.

- **Availability and Storage**: Fresh fenugreek leaves are available in most Indian markets and some specialty stores. They can be stored in the refrigerator for up to 3 days. Dried fenugreek leaves, also known as kasoori methi, are available in most Indian grocery stores and can be stored for up to 6 months.

In summary, fenugreek leaves are a nutritious and flavorful vegetable that can be used in a variety of Indian dishes. They offer numerous health benefits and can be easily incorporated into a healthy diet.

6.3.1. Agroclimatic Conditions Suitable for Growing Fenugreek Greens

Fenugreek leaves require specific agroclimatic conditions to grow well. Here are the ideal conditions:

- Fenugreek is a cool-season crop that thrives in temperate climates. It requires a moderate climate with warm days and cool nights. The ideal temperature for growing fenugreek leaves ranges from 15°C to 30°C.
- Fenugreek leaves require adequate moisture, especially during the germination and seedling stages. The crop requires consistent soil moisture, but excessive water can lead to root rot and other diseases. An annual rainfall of 600-800 mm is considered ideal for growing fenugreek leaves.
- Fenugreek can be grown on a variety of soils, but it prefers well-drained, fertile soils with a pH range of 6.0-7.0. The crop can tolerate some level of salinity, but extreme saline conditions can affect its growth.
- Fenugreek leaves require full sun to partial shade. The crop can be grown in areas with low to moderate levels of sunlight, but excessive shade can reduce its growth and yield.
- Fenugreek is typically grown as a rabi crop in India, which means it is sown in the winter months (October-November) and harvested in the spring months (February-March). In areas with mild winters, fenugreek can be grown as a kharif crop, which means it is sown in the summer months June and July. It is ready for leave 30 to 40 days after sowing.

6.3.2. Optimum Weather Conditions Required for High Productivity

For better vegetative growth of fenugreek leaves, the following optimum conditions are required:

- **Temperature**: The ideal temperature for vegetative growth of fenugreek leaves ranges from 20°C to 25°C. Temperatures above 30°C can lead to leaf scorch and reduced growth, while temperatures below 15°C can slow down growth.

- **Moisture:** Fenugreek leaves require consistent moisture, especially during the seedling stage. The soil should be kept moist but not waterlogged, as excessive water can lead to root rot and other diseases.
- **Light:** Fenugreek leaves require full sun to partial shade. Direct sunlight for 4-6 hours a day is ideal for promoting healthy growth.
- **Soil:** Fenugreek prefers well-drained, fertile soils with a pH range of 6.0-7.0. The soil should be rich in organic matter to support healthy microbial activity.
- **Nutrients:** Fenugreek leaves require adequate nutrients, especially nitrogen, phosphorus, and potassium. A balanced fertilizer application can promote healthy growth and development.
- **Sowing Density:** The ideal sowing density for fenugreek leaves is 10-15 kg/ha. This allows for proper aeration and sunlight penetration, promoting healthy growth and reducing the risk of disease.
- **Irrigation:** Fenugreek leaves require regular irrigation, especially during the seedling stage. The crop should be irrigated 10-15 days interval and may vary depending upon soil type and prevalent weather conditions. Sandy soils need more frequent irrigation..

By providing these optimum conditions, farmers can promote healthy vegetative growth and maximize the yield of fenugreek leaves.

6.3.3. Effect of Extreme Weather Conditions

Extreme weather conditions can significantly impact the vegetative growth of fenugreek leaves. Here's how different extreme weather conditions can affect the crop:

- **Drought**: Drought can severely impact the growth of fenugreek leaves. When the soil moisture is low, the plants may become stunted, and the leaves may become wilted and discolored. Prolonged drought can lead to reduced yields and lower quality leaves.
- **Excessive Rainfall:** Excessive rainfall can also negatively impact the growth of fenugreek leaves. Waterlogged soil can lead to root rot, which can cause the plants to become weak and susceptible to disease. Excessive rainfall can also lead to nutrient deficiencies, as the nutrients may be leached out of the soil.
- **High Temperatures**: High temperatures can cause heat stress in fenugreek plants, leading to reduced growth and yields. When the

temperature exceeds 35°C, the plants may become scorched, and the leaves may become discolored. Prolonged exposure to high temperatures can lead to reduced quality leaves and lower yields.

- **Low Temperatures**: Low temperatures can also impact the growth of fenugreek leaves. When the temperature drops below 10°C, the plants may become stunted, and the leaves may become discolored. Frost can also damage the leaves, leading to reduced yields and lower quality leaves.
- **Wind:** Strong winds can also impact the growth of fenugreek leaves. Wind can cause physical damage to the plants, leading to reduced growth and yields. Wind can also lead to soil erosion, which can reduce the fertility of the soil and impact the growth of the plants.
- **Hail:** Hail can cause significant damage to fenugreek leaves, leading to reduced yields and lower quality leaves. Hail can physically damage the leaves, causing them to become bruised and discolored. Prolonged exposure to hail can lead to reduced growth and yields.

By understanding the impact of extreme weather conditions on fenugreek leaves, farmers can take steps to mitigate these effects and promote healthy growth and development.

6.4. Vine Spinach/Malabar Spinach (*Basella alba*)

Malabar spinach is a tropical vine native to India and Southeast Asia. It is a popular leafy green vegetable in many Indian households, particularly in the southern and eastern regions (Fig. 6.5).

Fig. 6.5: Vine spinach crop

- **Area and Production in India**: According to the Indian Ministry of Agriculture and Farmers Welfare, Malabar spinach is grown in various parts of the country, with the following states being major producers:
    - Andhra Pradesh: Accounts for over 40% of India's Malabar spinach production.

    - Tamil Nadu: Accounts for around 20% of India's Malabar spinach production.
    - West Bengal: Accounts for around 15% of India's Malabar spinach production.
    - Odisha: Accounts for around 10% of India's Malabar spinach production.
- In terms of production, India produces over 1.5 million metric tons of Malabar spinach annually, with an estimated value of over ₹1,500 crores (approximately $200 million USD).
- Importance of Malabar Spinach in India: Malabar spinach is an important crop in India due to its:
    - Nutritional value: Rich in vitamins A, C, and K, as well as minerals like calcium and iron.
    - Cultural significance: Used in various traditional Indian dishes, such as dal, curries, and stir-fries.
    - Economic importance: Provides a source of income for many small-scale farmers and vegetable vendors.
- **Climate resilience**: Can thrive in hot and humid climates, making it an ideal crop for tropical regions.
- **Medicinal properties**: Used in traditional Indian medicine to treat various ailments, such as fever, cough, and digestive issues.

Overall, Malabar spinach is a valuable crop in India, providing nutritional, cultural, economic, and environmental benefits to the country.

6.4.1. Agroclimatic Conditions Suitable for Growing Vine Spinach

Vine spinach is a tropical plant that thrives in warm and humid climates. It requires specific agroclimatic conditions to grow optimally.

- Vine spinach grows best in regions with high temperatures, ranging from 25°C to 35°C. It can tolerate temperatures up to 40°C, but prolonged exposure to extreme heat can lead to reduced yields.
- In terms of rainfall, vine spinach requires adequate moisture, especially during the germination and seedling stages. It thrives in areas receiving annual rainfall of 600-1200 mm, with a relative humidity of 60-80%.
- Vine spinach can grow in a wide range of soils, but it prefers well-drained, fertile soils with a pH range of 6.0-7.5. It can tolerate some level of salinity, making it suitable for coastal regions.

- In India, vine spinach is typically grown during the summer and rainy seasons, from March to October. It is often cultivated as a monsoon crop, taking advantage of the abundant rainfall and high humidity during this period.

Overall, vine spinach is adapted to the warm and humid climates of tropical regions, making it an ideal crop for many parts of India.

6.4.2. Optimum Weather Conditions Required for High Productivity

Agroclimatic Conditions Suitable for High Productivity of Vine Spinach are:

- Vine spinach requires optimal agroclimatic conditions. The ideal temperature for vine spinach growth ranges from 25°C to 30°C. Temperatures above 35°C can lead to reduced yields and poor quality leaves.
- Adequate moisture is essential for vine spinach growth. It requires consistent and well-distributed rainfall of around 800-1000 mm per annum. Supplemental irrigation is necessary during periods of drought or low rainfall.
- Vine spinach grows best in areas with high relative humidity, ranging from 70% to 90%. This helps maintain leaf turgidity and promotes healthy growth.
- Soil with a pH range of 6.5-7.2 is ideal for vine spinach cultivation. The soil should be well-drained, fertile, and rich in organic matter.
- In terms of sunlight, vine spinach requires partial shade to full sun. Direct sunlight for 6-8 hours a day promotes healthy growth and development.
- In India, the ideal time for sowing vine spinach is during the spring-summer season (February-March) or the kharif season (June-July). This allows the crop to take advantage of the warm temperatures and abundant rainfall during these periods.

6.4.3. Effect of Extreme Weather Events

Vine spinach is a tropical crop that is sensitive to extreme weather events. These events can significantly impact the crop's growth, development, and productivity.

- **Drought:** Vine spinach is highly susceptible to drought, which can lead to reduced yields, poor leaf quality, and increased susceptibility to pests and diseases. Prolonged drought can cause the plant to wilt, reducing its ability to undergo photosynthesis and produce new growth.
- **Flooding:** Excessive rainfall and flooding can also negatively impact vine spinach. The crop is sensitive to waterlogging, which can cause the

roots to rot, leading to plant death. Flooding can also increase the risk of disease, as excess moisture creates an ideal environment for pathogens to develop.

- **Heatwaves:** Vine spinach is sensitive to extreme temperatures, particularly heatwaves. Prolonged exposure to temperatures above 40°C can cause the plant to suffer from heat stress, leading to reduced yields and poor leaf quality.
- **Cold snaps:** Vine spinach is also sensitive to cold temperatures. Frost can damage or kill the plant, while prolonged exposure to temperatures below 15°C can slow down growth and reduce yields.
- **Strong winds**: Strong winds can damage vine spinach plants, causing them to become tangled or uprooted. This can reduce yields and make the crop more susceptible to disease.

Overall, extreme weather events can have a significant impact on vine spinach, reducing yields, affecting quality, and increasing the risk of disease.

6.4.4. Sowing and Method of Propagation

Vine spinach is a tropical crop that can be propagated through seeds, cuttings, or layering.

- Sowing Seeds:
  - Seed selection: Choose high-quality seeds that are disease-free and suitable for the local climate.
  - Sowing time: In India, the ideal time for sowing vine spinach seeds is during the spring-summer season (February-March) or the kharif season (June-July).
  - Sowing depth: Sow seeds 1-2 cm deep in well-draining soil.
  - Sowing spacing: Sow seeds 10-15 cm apart, with rows spaced 30-40 cm apart.
  - Irrigation: Keep the soil consistently moist during the germination period (7-10 days).
- Propagation through Cuttings:
  - Cutting selection: Choose healthy, disease-free stem cuttings with at least two nodes.
  - Cutting length: Cut stems into 10-15 cm long sections.
  - Removal of lower leaves: Remove lower leaves, leaving only two or three sets of leaves at the top.

- Planting: Plant cuttings in well-draining soil, burying them up to the node.
- Irrigation: Keep the soil consistently moist during the rooting period (7-10 days).

- Propagation through Layering:
  - Layering: Bend a long stem of a mature vine spinach plant down to the ground, securing it with a rock or a U-shaped wire.
  - Rooting: Allow the stem to root for 7-10 days.
  - Cutting: Cut the stem below the node, leaving a small portion of stem attached to the roots.
  - Planting: Plant the layered stem in well-draining soil, burying it up to the node.
  - Irrigation: Keep the soil consistently moist during the establishment period.
- Vine spinach can be propagated throughout the year, but the ideal time is during the spring-summer season or the kharif season.
- Vine spinach, also known as Malabar spinach or Basella alba, typically takes around 50-60 days to mature from sowing the seeds. However, one can start harvesting the leaves in as little as 20-30 days after sowing, when they are young and tender.

6.5. Sorrel Leaves (*Rumex acetosa*)

- According to the Indian Ministry of Agriculture and Farmers Welfare, the area under sorrel leaf cultivation in India is not separately reported (Fig. 6.6). However, it is included under the category

Fig. 6.6: Sorrel leaves crop

- India is one of the largest producers of vegetables in the world.
- The country produced 188.94 million metric tons of vegetables in 2020-21.

Importance of Sorrel Leaves in India

Sorrel leaves, also known as "gongura" in Telugu and "pulicha keerai" in Tamil, are a popular leafy green vegetable in India, particularly in the southern states of Andhra Pradesh, Telangana, and Tamil Nadu.

- **Culinary uses**: Sorrel leaves are used in various dishes, such as curries, soups, and chutneys, due to their tangy and slightly sour taste.
- **Nutritional value**: Sorrel leaves are rich in vitamins A and C, iron, and antioxidants, making them a nutritious addition to the diet.
- **Economic importance:** Sorrel leaf cultivation provides livelihood support to many small-scale farmers in India, particularly in rural areas.
- **Cultural significance**: Sorrel leaves are used in traditional medicine and are also considered a sacred leaf in some Indian cultures.

Overall, sorrel leaves play a significant role in Indian cuisine, nutrition, and economy, particularly in the southern states.

6.5.1. Agroclimatic Conditions Required to Grow Sorrel Leaves

In regions with a distinct winter season, sorrel leaves can be sown in early winter (October-November) or late winter (January-February). In tropical regions, sorrel leaves can be sown throughout the year, except during the hottest summer months (May-June).

- Avoid sowing during heavy rainfall or waterlogged conditions, which can lead to seed rot and poor germination.
- Extreme temperatures, such as below 10°C or above 35°C, which can affect germination and seedling growth.
- Dry spells or drought conditions, which can impact seed germination and seedling establishment.

6.5.2: Optimum Weather Conditions Required for Better Vegetative Growth

For better vegetative growth of sorrel leaves, the optimum weather conditions are:

- **Temperature**: 15-25°C, with average temperatures. Temperatures above 30°C can lead to leaf scorch and reduced growth.
- **Relative Humidity**: 60-80%. High humidity promotes healthy growth, while low humidity can cause leaf drop.
- **Rainfall**: Well-distributed rainfall of 600-800 mm per annum. Consistent moisture levels promote healthy growth.

- **Sunlight**: Partial shade, especially during the hottest part of the day. Direct sunlight can cause leaf scorch and reduce growth.
- **Soil Moisture**: Consistent moisture levels in the soil, with a minimum of 50% relative soil moisture.
- **Wind:** Low wind speeds, as high winds can cause mechanical damage to the plants.
- **Duration**: The number of days required to harvest sorrel leaves after sowing can vary depending on factors like weather, soil quality, and the specific variety being grown. However, here are some general guidelines:
    - Baby leaves: 15-20 days after sowing, you can start harvesting baby sorrel leaves.
    - Mature leaves: 30-40 days after sowing, the plants will be ready for harvesting mature sorrel leaves.
    - Full maturity: 50-60 days after sowing, the plants will reach full maturity, and you can harvest the leaves, stems, and flowers.

6.6. Mint (*Mentha spicata*)

Mint leaves are a popular and versatile vegetable in Indian cuisine, particularly in North Indian and Pakistani cooking (Fig. 6.7). Here's an overview of mint leaves as a vegetable:

Fig. 6.7: Mint crop

- Mint leaves are used fresh or dried in a variety of dishes, such as curries, chutneys, salads, and drinks. They add a unique flavor and fragrance to food.
- In Indian cuisine, mint leaves are often used in combination with other herbs like cilantro, ginger, and garlic to create flavorful chutneys and sauces. Mint chutney, also known as "pudina chutney," is a popular condiment served with snacks, street food, and Indian meals.

- Mint leaves are also used in popular Indian dishes like biryani, pulao, and kebabs. They add a fresh and herbaceous flavor to these dishes.
- In addition to their culinary uses, mint leaves have several health benefits. They are rich in antioxidants, vitamins, and minerals, and are known to aid digestion, relieve nausea, and reduce inflammation. Some popular Indian dishes that use mint leaves as a vegetable include:
  - Mint chutney
  - Mint biryani
  - Mint pulao
  - Mint kebabs
  - Mint raita (a yogurt-based side dish)
  - Mint lassi (a yogurt-based drink)

Overall, mint leaves are a flavorful and nutritious vegetable that add depth and variety to Indian cuisine.

6.6.1. Agroclimatic Conditions Required for Growing Mint

Mint is a hardy, adaptable crop that can thrive in a variety of agroclimatic conditions. However, it grows best in certain environments. Here are the favorable agroclimatic conditions for growing mint:

- Mint grows best in temperate climates with moderate temperatures. The ideal temperature range is between 15°C and 30°C (59°F to 86°F). Temperatures above 35°C (95°F) can lead to leaf scorch and reduced oil quality.
- Mint prefers well-drained, fertile soils with a pH range of 6.0 to 7.0. It can grow in a variety of soil types, including clay, loam, and sand.
- Mint requires consistent moisture, especially during the germination and establishment phases. It thrives in areas with annual rainfall of at least 600 mm (24 in). However, excessive waterlogging can lead to root rot and other problems.
- Mint grows best in partial shade to full sun. Direct sunlight for 4-6 hours a day is ideal for promoting healthy growth and oil production.
- Mint can grow at a wide range of altitudes, from sea level to 1,500 meters (4,921 ft) above sea level.
- Mint prefers a relatively high relative humidity of 60-80%. This helps maintain leaf turgidity and promotes healthy growth.
- In India, the major mint-growing regions have the following agroclimatic conditions:

- The Indo-Gangetic Plain, which includes states like Uttar Pradesh, Punjab, and Haryana, has a subtropical climate with hot summers and cool winters. This region is ideal for mint cultivation.
- The Himalayan foothills, which include states like Uttarakhand and Himachal Pradesh, have a temperate climate with mild winters and cool summers. This region is also suitable for mint cultivation. Extreme weather conditions can significantly impact mint crops.

6.6.2. Effect of Extreme Weather Conditions

Droughts and Heat Waves can lead to water stress, causing mint leaves to become brittle and develop brown tips. On the other hand, Excessive Rainfall and Flooding can result in root rot and other fungal diseases, damaging the crop.

6.7. Kale (*Brassica oleracea* L. var. acephala)

Kale is a cool-season crop that is a member of the cabbage family and a healthy leafy green vegetable. It is one of the most popular and nutritious vegetables around the world and its popularity are steadily increasing in India (Fig. 6.8). The Scientific name of Kale is *Brassica oleracea*; it belongs to the same species of Cabbage, Broccoli, and Cauliflower, but it does not form any heads.

Fig. 6.8: Kale crop

- **Regions**: Kale cultivation is concentrated in the northern states of India that have suitable agro-climatic conditions. Himachal Pradesh, Uttar Pradesh, Jammu and Kashmir, Punjab, Haryana, and the Nilgiris hills are known for kale cultivation.

6.7.1. Agroclimatic Conditions Required to Grow Kale

Kale thrives in cold and dry climates, making it a winter crop and seeds sown in late October or early November. Kale is very hardy crop and tolerates frost and cold temperatures. Some of the varieties kale can with temperatures up to -15^0 C.

It prefers well drained soil rich in organic matter. It can be grown by either direct seeding or transplantation of seedlings. Mulching is better to maintain soil temperature.

6.8. Lettuce (*Lactuca sativa* L.)

India ranks 3rd in commercial production of lettuce and occupy 4% of the World's total production. The Nilgiri mountain of South India is emerging as lettuce production regions (Fig. 6.9).

Fig. 6.9: Lettuce crop

The demand for lettuce is growing very fast. The restaurant industry is booming steadily because the country's young generation preferring international cuisine like a burger, pizza, sandwiches, tacos, spring rolls, etc., where lettuce is an integral part.

- **Lettuce Variety:** Among the cultivated varieties, Iceberg and Great Lakes are the popular head type lettuce grown in India. The head of iceberg lettuce is light green, round in shape, and tightly packed with lettuce leaves like cabbage head. Lettuce is an excellent source of different vitamins and minerals. Iceberg lettuce has high water content (95 percent) and low in calories (13 kcal/100g), which is very effective for weight loss efforts. Iceberg lettuce helps in fighting heart diseases and strokes and protecting infants from neural tube defects during pregnancy.
- Regular inclusion of lettuce in salads is known to prevent osteoporosis, iron-deficiency anemia and believed to protect from cardiovascular diseases, age-related macular disease (ARMD) in the elderly, Alzheimer's disease and cancers.

6.8.1. Agroclimatic Requirements for Lettuce Cultivation

Lettuce is a cool season crop and grows well when the mean daily temperatures are in he range 10 to 20 degreeC. Cold nighst are congenial for better quality. Regions with mild summers and mild winter .Well-Drained fertile soil with a pH of about 6.0 is most suitable for the growth of lettuce. It is moderately salt tolerant.

6.9. Collard Greens (*Brassica oleracea* var. viridis)

- Collard greens are a leafy green in the cabbage family. These greens stand out from other cabbage-family plants in their tolerance to both cold and hot temperatures, as well as high humidity. Their loose-leaf (non-heading) form makes them less susceptible to fungal diseases compared with their tight-leaved and head-forming relatives, such as cabbage and cauliflower (Fig. 6.10). Cold-hardy and heat-tolerant, collards grow well in all regions and are suitable for planting in the garden, in raised beds, and in containers.

Fig. 6.10: Collard greens crop

Although the leaves can be harvested at any time, their flavor becomes sweeter after a few frosts. In northern climates, collards are usually grown for a fall harvest. In India, collard greens, locally known as "haakh," are primarily grown and consumed in the Kashmir Valley and are a staple in Kashmiri cuisine, particularly during the winter months.

6.10. Water Cress (*Nasturtium officinale*)

- Water cress is a native water or semiaquatic plant that has a high nutrient density. It is a popular leafy green vegetable which is traditionally served cooked although it is increasingly being consumed raw in salads, sandwiches or as a garnish. Water cress is a perennial aquatic or semi-aquatic plant species native to Europe and Asia. Watercress grows in nutrient-rich, streaming freshwater (Fig. 6.11).

Fig. 6.11: Water cress crop

6.10.1. Agroclimatic Requirements

Watercress is traditionally used as winter salad as it grows in flowing water even at cool temperatures as long as the water is not frozen. Usually, watercress is cultivated in sophisticated held back streaming waters, but also grows well in moist soils.

Climate and Soil: Water cress grows as a cool season crop. The ideal temperature for growth is 13- 24^{0}C. The crop can be grown in all types of soil.

6.10.2. Harvesting and Yield

Harvestable watercress plants can be produced in about 35 days from sowing Harvest the dark green leaves at any time during the year. Leaves and young stems taste best if harvested before watercress flowers. Normally the plants are harvested when they reach a height of 17. 8 cm (7 inches) and subsequent harvesting is done at 15days intervals. The average annual yield of watercress is 6-8tonnes/ha.

- **Post Harvest Management**: Water cress requires rapid and careful post-harvest handling to maintain quality. For best quality maintenance watercress should be kept at 00C and 95% relative humidity (RH) or above throughout storage and marketing. Watercress bunched in polyethylene bags and at 00C and 95% RH remained marketable for up to 4 weeks.
- Watercress is a native water or semiaquatic plant that has a high nutrient density. It is a popular leafy green vegetable which is traditionally served cooked although it is increasingly being consumed raw in salads, sandwiches or as a garnish.

6.11. Curry Leaves (*Murraya koenigii*)

Curry leaves are a staple herb in Indian cuisine, particularly in southern and western regions. Native to India and Sri Lanka, curry leaves are a key ingredient in various traditional dishes, including curries, chutneys, & pickles (Fig. 6.12).

Fig. 6.12: Curry leaves plant

- India is the largest producer of curry leaves, accounting for over 70% of global production. The major curry leaf-producing states in India are Tamil Nadu, Karnataka, Kerala, Andhra Pradesh, and Telangana. In terms of production, India produces over 1.5 million tons of curry leaves annually, with a value of over ₹1,500 crores (approximately $200 million USD).
- Curry leaves hold significant cultural, economic, and medicinal value in India. They are an essential ingredient in various Indian dishes, adding flavor, aroma, and texture. In traditional Indian medicine (Ayurveda), curry leaves have been used for centuries for their anti-inflammatory, antioxidant, and antibacterial properties. As a significant cash crop, curry leaves provide a source of income and employment for many farmers in India.
- Furthermore, curry leaves are an integral part of Indian culture and tradition. They are often used in religious ceremonies and are considered a symbol of good luck and prosperity. Overall, curry leaves play a vital role in Indian cuisine, culture, and economy, extending beyond their culinary use to offer medicinal and economic benefits.

6.11.1. Agroclimatic Requirement

Curry leaf is a tropical plant that thrives in warm and humid climates.

- To cultivate curry leaves, a temperature range of 15°C to 35°C is ideal, as they are sensitive to frost and temperatures below 10°C can damage the plant.
- Adequate rainfall is also essential, with a minimum annual rainfall of 600 mm. However, curry leaf plants can tolerate a wide range of rainfall patterns. The plants prefer well-draining soil with a pH range of 6.0 to 7.0 and can grow in various soil types, including clay, loam, and sand.
- Humidity is another crucial factor, with a relative humidity of 60% to 80% being ideal for curry leaf growth. In terms of sunlight, the plants

require partial shade to full sun, depending on the climate. In warmer climates, they prefer partial shade, while in cooler climates, they can tolerate full sun.

Overall, curry leaf cultivation is suitable for tropical and subtropical regions with mild winters and hot summers. The plants thrive in warm and humid climates with adequate rainfall and well-draining soil, making them an ideal crop for many Indian states.

6.12. Coriander Leaves (*Coriandrum sativum* L.)

Coriander leaves are an essential herb in Indian cuisine, particularly in North Indian and South Indian cooking (Fig. 6.13). India is one of the largest producers and consumers of coriander leaves in the world.

Fig. 6.13: Coriander leaves crop

- The major coriander leaf-producing states in India are Rajasthan, Gujarat, Madhya Pradesh, Uttar Pradesh, and Tamil Nadu. According to the Indian Spice Board, India produces over 4.5 million tons of coriander leaves annually, with a value of over ₹3,500 crores (approximately $450 million USD).
- Coriander leaves hold significant cultural, culinary, and medicinal value in India. They are a key ingredient in various Indian dishes, including curries, chutneys, salads, and street food, adding flavor, aroma, and freshness. In traditional Indian medicine (Ayurveda), coriander leaves have been used for centuries for their anti-inflammatory, antioxidant, and antibacterial properties.
- In addition to their culinary and medicinal significance, coriander leaves are also an integral part of Indian culture and tradition. They are often used in religious ceremonies and are considered a symbol of good luck and prosperity. As a significant cash crop, coriander leaves provide a source of income and employment for many farmers in India. Overall, coriander leaves play a vital role in Indian cuisine, culture, and economy.

6.12.1. Agroclimatic Requirement

Coriander leaves require specific agroclimatic conditions to thrive. Here are the ideal conditions for coriander leaves production:

- Coriander leaves prefer a cool and dry climate, with temperatures ranging from 15°C to 25°C. They are sensitive to high temperatures and humidity, which can lead to bolting (premature flowering) and reduced yields.
- In terms of rainfall, coriander leaves require adequate moisture, especially during the germination and seedling stages. An annual rainfall of 600-800 mm is ideal, with supplemental irrigation provided during periods of drought.
- Coriander leaves can grow in a variety of soils, but well-draining sandy loam soils with a pH range of 6.0-7.0 are ideal.
- Full sun to partial shade is suitable for coriander leaves production. However, in warmer climates, providing some shade can help prevent bolting.

Overall, coriander leaves production is best suited for regions with mild winters, cool summers, and well-defined wet and dry seasons.

6.13. Conclusion

- The agroclimatology of greens plays a vital role in their cultivation and production. Understanding the specific agroclimatic requirements of various green leafy vegetables, such as curry leaves and coriander leaves, is essential for optimizing their growth, yield, and quality. The ideal temperature, rainfall, soil, and sunlight conditions for each type of green vary, and even slight deviations from these conditions can significantly impact their production.
- By recognizing the agroclimatic requirements of greens, farmers and agricultural practitioners can adopt strategies to mitigate the adverse effects of climate variability and change. This includes using irrigation management techniques, mulching, and other conservation practices to optimize water use and reduce soil temperature. Furthermore, understanding the agroclimatology of greens can inform breeding programs aimed at developing climate-resilient varieties.
- Ultimately, the agroclimatology of greens is critical for ensuring the long-term sustainability and productivity of these essential crops. By applying agroclimatic principles and practices, we can promote the efficient use of natural resources, reduce the environmental impact of green production, and contribute to food security and sustainable agriculture.

7

Agroclimatology of Allium Vegetables

Allium vegetables, which include onions, garlic, and shallots, are a fundamental part of Indian cuisine and play a vital role in the country's agricultural sector (Fig. 7.1). Here's an overview of their importance, area, and production in India:

- India is the second-largest producer of onions in the world, accounting for over 20% of global production. Onions are grown across various states, with Maharashtra, Madhya Pradesh, and Karnataka being the top producers. The country produces over 19 million tons of onions annually, with a significant portion being consumed domestically.
- Garlic is another essential Allium crop in India, with the country producing around 1.4 million tons annually. The major garlic-producing states in India are Madhya Pradesh, Gujarat, and Maharashtra.
- Shallots are also grown in India, particularly in the states of Maharashtra, Gujarat, and Tamil Nadu. The country produces around 0.7 million tons of shallots annually.
- The importance of Allium vegetables in India extends beyond their culinary uses. They are a significant source of income for millions of farmers across the country. The Allium crop industry also provides employment opportunities in rural areas, contributing to the country's economic growth.
- In terms of area, Allium vegetables are grown across over 1.3 million hectares of land in India. The majority of this area is dedicated to onion cultivation, followed by garlic and shallots. Onions and garlic are often referred to as root vegetables, but technically, they are not roots. They belong to the Allium genus and are classified as bulbous vegetables.

Fig. 7.1: Allium vegetables

The reason they are sometimes considered root vegetables is due to the following reasons:

- Underground growth: Onions and garlic grow underground, similar to roots. They form bulbs or cloves beneath the soil surface.
- Storage organs: Like roots, onions and garlic have storage organs that hold nutrients and water. In the case of onions and garlic, these storage organs are the bulbs or cloves.
- Culinary and cultural associations: In many cuisines, especially in Asian and European cooking, onions and garlic are used in similar ways to root vegetables like carrots, beets, and turnips. They're often roasted, sautéed, or boiled, which reinforces the association with root vegetables.
- While onions and garlic don't fit the botanical definition of root vegetables, their underground growth, storage organs, and culinary uses have led to their common classification as root vegetables in a broader, more general sense.

Overall, Allium vegetables are a vital part of India's agricultural sector, and their production plays a significant role in the country's food security and economy. As garlic and onions are already covered under root vegetables already, the information available on other allium vegetables is provided in this chapter.

7.1. Shallots (*Allium cepa* gr.aggregatum)

Shallots are a type of Allium vegetable that belongs to the same family as onions and garlic. They are characterized by their smaller size, sweeter flavor, and milder aroma compared to onions (Fig. 7.2).

Fig. 7.2: Shallots crop

Shallots have a distinctive coppery or reddish-brown color and are often used in cooking for their unique flavor and texture.

- In India, shallots are grown across various states, with Maharashtra, Gujarat, and Tamil Nadu being the top producers. The country produces around 0.7 million tons of shallots annually, with a significant portion being consumed domestically.
- Shallots are an important crop in India, not only for their culinary uses but also for their economic significance. They provide a source of income for thousands of farmers, particularly small-scale and marginal farmers, who cultivate shallots in their fields. The shallot crop also generates employment opportunities in rural areas, contributing to the country's economic growth.
- In terms of area, shallots are grown across over 0.2 million hectares of land in India. The majority of this area is dedicated to small-scale farming, with farmers often cultivating shallots as a subsidiary crop to their main crop.

The shallots are a vital crop in India, and their production plays a significant role in the country's food security, economy, and rural development. Shallots are a cool-season crop that thrives in temperate climates with moderate temperatures and adequate moisture.

7.1.1. Agroclimatic Conditions Required to Grow Shallots

The ideal agroclimatic conditions for growing shallots are characterized by:

- Temperature is a critical factor in shallot cultivation. Shallots require a relatively cool temperature during the growing season, with optimal temperatures ranging from 15°C to 25°C. Temperatures above 30°C can lead to bolting, while temperatures below 10°C can slow down growth.
- Shallots require adequate moisture, especially during the germination and bulb formation stages. The crop requires a well-distributed annual

rainfall of around 600-800 mm. However, excessive rainfall can lead to waterlogged soil conditions, which can be detrimental to the crop.

- Shallots prefer full sun to partial shade, with a minimum of 6 hours of direct sunlight per day. However, in warmer climates, providing some shade during the hottest part of the day can help prevent scorching of the leaves.
- Soil is another critical factor in shallots cultivation. Shallots prefer well-draining, fertile soils with a pH range of 6.0-7.0. The ideal soil texture is a mix of sand, silt, and clay, with good organic matter content.
- In terms of altitude, shallots can be grown at a wide range of elevations, from sea level to around 1,500 meters above sea level. However, at higher elevations, the crop may require additional protection from frost and cold temperatures.
- The ideal time for planting shallots varies depending on the region and climate. In general, shallots are planted in the fall or early spring, when the weather is cool and there is adequate moisture. In regions with mild winters, shallots can be planted in the late winter to early spring.

Overall, shallots require a specific set of agroclimatic conditions to thrive. By providing the right combination of temperature, moisture, sunlight, soil, and altitude, farmers can optimize shallot yields and quality.

7.1.2. Optimum Weather Conditions Required for Sowing

The optimum weather conditions required for sowing shallots are characterized by:

- **Cool temperatures**: Shallots thrive in cool temperatures, with ideal temperatures ranging from 10°C to 20°C. Temperatures above 25°C can lead to bolting, while temperatures below 5°C can slow down germination.
- **Moderate humidity**: Shallots require a relatively high humidity during the germination phase, typically above 60%. However, excessive humidity can lead to rot and other diseases.
- **Adequate moisture**: Shallots need consistent moisture during the germination and seedling stages. The soil should be kept moist but not waterlogged, with a minimum of 50 mm of rainfall or irrigation per month.
- **Low wind speeds**: Shallots are susceptible to wind damage, especially during the seedling stage. Wind speeds above 15 km/h can cause damage to the plants.

- Overcast skies: Shallots prefer partial shade, especially during the hottest part of the day. Overcast skies can provide the ideal conditions for germination and seedling growth.
- **Soil temperature**: The ideal soil temperature for sowing shallots is between 10°C and 15°C. This allows for optimal germination and seedling growth.

In general, the ideal time for sowing shallots is during the cooler months of the year, typically from September to November or from February to April, depending on the region and climate.

- **Major Production Regions**: Tamil Nadu and Karnataka are the primary states for shallots production in India, with around 90% of the country's output coming from these two states. In Tamilnadu, shallots are grown inPerambalur, Ariyalur, Trichy, Dindigul, Namakkal, Erode, Tenkasi and Coimbatore districts. Shallots are a type of onion that belongs to the Allium family. They have a shorter shelf life compared to onions, but proper storage and weather conditions can help extend their storage life.
- **Optimal Storage Conditions:** To maximize the shelf life of shallots, follow these storage guidelines:
  - Store shallots in a cool, dry place with good ventilation. The ideal storage temperature is between 32°F (0°C) and 50°F (10°C). Maintain a relative humidity of 60-70%, as high humidity can lead to mold and rot. Store shallots in a dark place or use a breathable bag to block out light. Ensure good air circulation to prevent moisture buildup.
  - **Shelf Life:** The shelf life of shallots depends on the storage conditions and the freshness of the shallots when stored:
  - Whole shallots can last 1-2 months at room temperature, 2-4 months in the refrigerator, and 4-6 months in a cool, dark place. Chopped or minced shallots typically last 1-2 weeks in the refrigerator and up to 6 months in the freezer.
  - **Weather Conditions for Storage**: Avoid storing shallots in areas with direct sunlight, which can cause shallots to sprout or become bitter. High temperatures above 60°F (15°C) can lead to spoilage. High humidity can cause mold and rot, while flooding or waterlogging can cause shallots to rot.
  - **Tips for Longer Storage**: Shallots are curd before storing by drying them in a warm, dry place for 7-10 days to reduce moisture content. The breathable containers such as paper bags, mesh bags, or ventilated containers are used to store shallots to ensure they are not spoiled.

7.2. Leeks (*Allium ampeloprasum*)

Leeks are a type of vegetable that belongs to the Allium family, which also includes onions, garlic, and shallots. They are a cool-season crop, characterized by their mild onion flavor and crunchy texture (Fig. 7.3).

Fig. 7.3: Leek crop

- According to the Indian government's data, leeks are grown in various parts of the country, with major production areas including Himachal Pradesh, Jammu and Kashmir, and Uttarakhand. Himachal Pradesh is known for its favorable climate and soil conditions, making it a significant producer of leeks in India.
- The state's temperate climate and well-drained soil make it an ideal place for leek cultivation. Leeks are grown in various districts of Uttarakhand, including Dehradun, Nainital, and Pauri Garhwal. Leeks are a nutritious vegetable, rich in vitamins A, C, and K, as well as minerals like potassium and manganese.
- They are a versatile ingredient and can be used in soups, stews, salads, and as a flavoring agent in various dishes.
- Leeks have been traditionally used to treat various health issues, including digestive problems and respiratory infections.
- Leek cultivation provides a source of income for many farmers in India, particularly in the Himalayan region.
- Leeks are an integral part of various traditional cuisines in India, including Kashmiri and Himachali cuisine.

7.2.1. Agroclmatic Conditions Required

Leeks are a cool-season crop that thrive in temperate climates with moderate temperatures and adequate moisture. Here are the agroclimatic conditions required for growing leeks:

- Leeks require a long, cool growing season with temperatures ranging from 15°C to 25°C. Optimal growth occurs in temperatures between 18°C and 22°C.
- Leeks need adequate moisture, especially during the germination and seedling stages. A consistent supply of moisture ensures healthy growth and development. Aim for a well-distributed annual rainfall of at least 600 mm.
- Leeks prefer full sun to partial shade. Direct sunlight promotes healthy growth, but excessive heat can lead to bolting (premature flowering).
- Leeks thrive in well-draining, fertile soils with a pH between 6.0 and 7.0. A mix of sand, silt, and clay provides optimal growth conditions.
- Adequate vernalization (exposure to cold temperatures) is necessary for leeks to form bulbs. A period of 30 to 60 days with temperatures below 10°C helps initiate bulbing.

7.2.2. Ideal Weather Conditions for Sowing

The ideal weather conditions required for sowing leeks include a temperature range of 10-20°C (50-68°F), which allows for optimal germination and growth.

- Consistent and moderate moisture levels, with a relative humidity of 60-80%, are also essential.
- The soil temperature should be around 7-10°C (45-50°F) for optimal germination. Leeks prefer partial shade to full sun, but protection from extreme heat and direct sunlight during the hottest part of the day is necessary.
- Low to moderate wind speeds are also desirable, avoiding strong winds that can dislodge seeds or damage young seedlings.
- Finally, well-distributed rainfall or supplemental irrigation, providing about 20-30 mm of moisture per week, is necessary for healthy growth.

7.2.3. Shelf life and Weather Conditions Required for Storage

Leeks have a relatively short shelf life and require specific weather conditions for storage.

- The ideal storage conditions for leeks include a cool, dry, and well-ventilated area with a temperature range of 0-5°C (32-41°F). The relative humidity should be around 80-90% to maintain moisture levels and prevent drying out.
- It's essential to store leeks in a dark place to prevent greening and the development of a bitter flavor. Under these conditions, leeks can

be stored for around 2-3 months. If stored at room temperature, leeks typically last for 1-2 weeks.

7.3. Garlic Scapes (*Allium sativum* L.)

- Garlic scapes are the curly, green shoots that emerge from the center of hardneck garlic plants, and they are harvested to encourage the plant to focus its energy on bulb development rather than producing flowers and seeds (Fig. 7.4). Whole garlic scape is edible, including the flower bulb at the end. Some people remove the bulb because it has a thinner texture than the rest of the stalk.
- It can taste fibrous when consumed raw and will cook at a different rate when sauteing or roasting scapes.

Fig. 7.4: Garlic scapes crop

7.4. Scallions (*Allium fistulosum*)

Scallions, also known as green onions or spring onions, are a type of vegetable that belongs to the Allium family (Fig. 7.5). They are essentially immature onions that are harvested before the bulb forms.

- Scallions have a mild onion flavor and are characterized by their long thin stems with a small white or yellowish base with green leaves that are hollow and tube-like with mild sweet flavor that is often used to add depth to various dishes
- Scallions are a popular ingredient in many cuisines, particularly in Asian and Latin American cooking. They're often used as a garnish, added to salads, or used as a flavoring agent in soups, sauces, and marinades. Some popular uses for scallions include:
  - Garnish for soups, salads, and sandwiches
  - Added flavor in stir-fries, sautés, and grilled dishes

- Used as a topping for tacos, grilled meats, and vegetables
- Added to salsas, sauces, and marinades for extra

Fig. 7.5: Scallions crop

- Scallions (also known as green onions or spring onions) are a popular crop in many parts of the country, particularly in the states of:
  - Maharashtra: Known for its high production of scallions, with major growing areas in Nashik, Pune, and Satara districts.
  - Gujarat: Scallions are grown mainly in the districts of Ahmedabad, Vadodara, and Surat.
  - Uttar Pradesh: Scallion cultivation is common in the districts of Lucknow, Kanpur, and Varanasi.
  - Karnataka: Scallions are grown in the districts of Bangalore, Mysore, and Hassan.
- Scallions are typically grown in well-drained soil and require adequate moisture, especially during the germination and bulbing stages. In India, scallions are usually harvested within 60 to 90 days of sowing, depending on factors like weather, soil quality, and variety.

7.5. Conclusion

Allium vegetables, including onions, garlic, leeks, chives, and scallions, play a vital role in global cuisine and nutrition. These vegetables are not only a fundamental flavor component in various dishes but also offer numerous health benefits due to their richness in vitamins, minerals, and antioxidants.

- The cultivation of Allium vegetables is widespread, with different regions specializing in specific types. India, for instance, is a significant producer of onions and garlic, while other countries like China and the United States are major producers of various Allium crops.
- The versatility of Allium vegetables is evident in their various uses, from cooking and medicine to cosmetics and spiritual practices.

Furthermore, research continues to uncover the potential health benefits of these vegetables, including their anti-inflammatory, antibacterial, and anticancer properties.

- As the global population grows, the demand for Allium vegetables is likely to increase, driving innovation in their cultivation, processing, and utilization.
- By embracing sustainable agricultural practices and exploring new applications for Allium vegetables, we can ensure a steady supply of these essential crops while promoting environmental stewardship and human well-being.

8

Agroclimatology of Brassica Group Vegetables

Brassica is a genus of plants belonging to the mustard family (Brassicaceae) and includes some of the most widely cultivated and consumed vegetables globally. The Brassica group encompasses a diverse range of cool-season crops, including broccoli, cauliflower, cabbage, kale, and mustard, among others (Fig. 8.1). These vegetables are not only rich in nutrients, fiber, and antioxidants but also play a vital role in human health, food security, and sustainable agriculture.

Fig. 8.1: Brassica group vegetables

- India is one of the largest producers of Brassica vegetables, with a significant portion of the crop being cultivated in the Indo-Gangetic plains. According to the Indian Horticultural Database, Brassica vegetables are grown on approximately 2.5 million hectares of land, with a total production of over 30 million metric tons annually. The major Brassica-producing states in India include Maharashtra, Madhya Pradesh, Gujarat, Uttar Pradesh, and West Bengal.
- The importance of Brassica vegetables extends beyond their nutritional value and economic significance. They are also an integral part of

Indian cuisine and culture, with various regional and traditional recipes featuring these vegetables as main ingredients. Furthermore, Brassica crops are a vital component of crop rotation and intercropping systems, helping to maintain soil health, reduce pests and diseases, and promote biodiversity.

In the context of agroclimatology, understanding the climatic requirements and constraints of Brassica vegetables is crucial for optimizing their growth, yield, and quality. This chapter will delve into the agroclimatic aspects of Brassica vegetables, including their temperature, moisture, and solar radiation requirements, as well as the impact of climate change on their production and sustainability.

8.1. Cauliflower (*Brassica oleracea var. botrytis* L.)

Cauliflower is one of the most widely cultivated and consumed vegetables in India, belonging to the Brassica family (Fig. 8.2). It is a cool-season crop that thrives in the temperate climate of the Indian subcontinent. India is the second-largest producer of cauliflower globally, accounting for over 10% of the world's total production.

Fig. 8.2: Cauliflower crop

- **Area under Cauliflower Cultivation**: Cauliflower is grown in almost all states of India, with the major producing states being Maharashtra, Madhya Pradesh, Gujarat, Uttar Pradesh, and West Bengal. According to the Indian Horticultural Database, the area under cauliflower cultivation in India has been steadily increasing over the years, from approximately 0.45 million hectares in 2001-02 to around 0.65 million hectares in 2020-21. This expansion in area can be attributed to the growing demand for cauliflower in both domestic and international markets.
- **Production of Cauliflower**: India's cauliflower production has also shown a significant increase over the years, driven by improved crop

yields and expansion in cultivated area. During 2020-21, India produced around 7.5 million metric tons of cauliflower, with an average yield of 11.5 metric tons per hectare. The major cauliflower-producing states in India, such as Maharashtra and Madhya Pradesh, have achieved even higher yields, averaging around 15-16 metric tons per hectare.

- **Importance of Cauliflower:** Cauliflower is an important vegetable crop in India, playing a vital role in the country's food security and economy. Its importance can be gauged from the following aspects:
  - Cauliflower is a rich source of essential nutrients like vitamins, minerals, and antioxidants, making it a popular ingredient in various Indian cuisines. It is also used in traditional medicine, particularly in Ayurveda and Unani systems.
  - The cauliflower crop provides employment opportunities to millions of farmers, laborers, and traders across the country. It is estimated that the cauliflower industry supports the livelihoods of over 2 million people in India.
  - Cauliflower is also an important export commodity for India, with the country exporting significant quantities to countries like the United Arab Emirates, Saudi Arabia, and the United Kingdom. This generates substantial foreign exchange earnings for the country.

8.1.1. Agroclimatic Conditions Suitable for Growing Cauliflower

Cauliflower is a cool-season crop that requires specific agroclimatic conditions to grow. Here's a detailed overview:

- Temperature is a critical factor in cauliflower cultivation. The ideal temperature for cauliflower growth ranges from 15°C to 25°C. It requires a period of cooler temperatures, around 10°C to 15°C, to induce curd formation. Temperatures above 25°C can lead to curd discoloration and reduced quality.
- Cauliflower requires a moderate level of rainfall, ranging from 500 mm to 700 mm per annum. However, it is sensitive to excessive rainfall, which can lead to waterlogged soil and root rot. Well-distributed rainfall during the growing season is essential for optimal growth.
- Cauliflower prefers a cool and humid climate, with relative humidity ranging from 60% to 80%. Low humidity can lead to curd discoloration and reduced quality.
- Soil requirements for cauliflower are specific. It prefers well-draining, fertile soils with a pH range of 6.0 to 7.0. Cauliflower is sensitive to acidic soils, and alkaline soils can lead to nutrient deficiencies.

- Cauliflower is typically grown in regions with low to moderate wind speeds. High winds can cause physical damage to the plants and reduce curd quality.

In India, the ideal regions for cauliflower cultivation are the plains of North India, the Nilgiri hills, and the Western Ghats. These regions provide the necessary cool temperatures, moderate rainfall, and fertile soils for optimal cauliflower growth.

8.1.2. Optimum Weather Conditions Required for Sowing

For optimal sowing conditions, cauliflower requires a specific set of weather conditions:

- The ideal time for sowing cauliflower is during the cooler months of the year, when the temperature is around 15°C to 20°C. This allows for optimal germination and seedling establishment.
- Soil temperature is also crucial for sowing cauliflower. The ideal soil temperature for sowing is between 10°C to 15°C. This temperature range promotes healthy root development and reduces the risk of seed rot.
- Adequate moisture in the soil is essential for sowing cauliflower. The soil should be kept consistently moist during the germination period, which typically lasts around 7-10 days.
- Weather conditions with low wind speeds and reduced evapotranspiration rates are favorable for sowing cauliflower. This helps to prevent soil moisture loss and ensures optimal seed germination.
- In regions with distinct seasonal patterns, the best time for sowing cauliflower is typically:
    - In the plains of North India: September-October or January-February
    - In the hills of North India: April-May or August-September
    - In South India: August-September or December-January

Sowing cauliflower during these periods allows farmers to take advantage of the optimal weather conditions, resulting in healthier seedlings and higher yields.

8.1.3. Weather Conditions Favorable for High Productivity

For high production of cauliflower, specific weather conditions are required during the crop's growth stages:

- During the vegetative growth stage, cauliflower requires moderate temperatures ranging from 15°C to 20°C. This temperature range promotes healthy leaf growth and development.

- As the crop approaches the reproductive stage, a period of cooler temperatures, around 10°C to 15°C, is essential for curd formation and development. Temperatures above 20°C can lead to premature flowering and reduced curd quality.
- Adequate moisture is crucial for cauliflower production. The crop requires consistent and well-distributed rainfall, ranging from 500 mm to 700 mm per annum. Soil moisture level should be maintained ranging from 60% to 80 % of the soil's water-holding capacity.
- Relative humidity plays a significant role in cauliflower production. A relative humidity of 60% to 80% is ideal for promoting healthy growth and curd development.
- Weather conditions with low wind speeds and reduced evapotranspiration rates are favorable for cauliflower production. This helps to prevent moisture loss and reduces the risk of crop damage.
- In terms of sunlight, cauliflower requires partial shade to full sun, depending on the temperature. In warmer climates, partial shade can help to prevent scorching of the curd.

Overall, the ideal weather conditions for high cauliflower production include moderate temperatures, adequate moisture, and moderate humidity, with minimal wind and evapotranspiration.

8.1.4. Effect of Extreme Weather Events

Extreme weather events can have a significant impact on cauliflower crops, affecting their growth, development, and overall yield.

- Drought is one of the most significant extreme weather events affecting cauliflower production. Prolonged drought can lead to reduced plant growth, lower curd quality, and decreased yields. Cauliflower requires consistent moisture, especially during the curd formation stage, and drought can disrupt this process.
- On the other hand, excessive rainfall can also be detrimental to cauliflower crops. Waterlogged soil can lead to root rot, reducing plant growth and increasing the risk of disease. Excessive rainfall can also cause curd discoloration and reduce its quality.
- High temperatures are another extreme weather event that can impact cauliflower production. Temperatures above 25°C can lead to premature flowering, reduced curd quality, and lower yields. Cauliflower is a cool-season crop and requires moderate temperatures to thrive.

- Low temperatures can also affect cauliflower crops, especially if they occur during the curd formation stage. Frost can damage or destroy the curd, reducing yields and affecting quality.
- Windstorms can also impact cauliflower production, especially if they occur during the curd formation stage. Strong winds can cause physical damage to the plants, reducing yields and affecting quality.
- Hailstorms can also be devastating to cauliflower crops, causing physical damage to the plants and reducing yields.
- In addition to these extreme weather events, climate change is also affecting cauliflower production. Changes in temperature and precipitation patterns are altering the growing seasons and affecting the crop's growth and development.
- To mitigate the effects of extreme weather events on cauliflower production, farmers can use various strategies such as:
  - Mulching to conserve soil moisture and reduce soil temperature
  - Using drip irrigation to reduce water loss and ensure consistent moisture
  - Planting cauliflower varieties that are resistant to extreme weather events
  - Using row covers to protect the plants from wind, hail, and frost
  - Implementing conservation agriculture practices to reduce soil erosion and improve soil health

By using these strategies, farmers can reduce the risks associated with extreme weather events and improve the resilience of their cauliflower crops.

8.2. Cabbage (*Brassica oleracea* var. capitata)

Cabbage is one of the most widely cultivated and consumed vegetables in India. It is a cool-season crop that belongs to the Brassica family (Fig. 8.3).

- In terms of area, cabbage is grown in almost all parts of India, with the major producing states being Maharashtra, Gujarat, Madhya Pradesh, Uttar Pradesh, and West Bengal. According to the data from the Ministry of Agriculture and Farmers Welfare, the total area under cabbage cultivation in India was around 397,000 hectares in 2020-21.
- Cabbage production in India has been steadily increasing over the years. The total production of cabbage in India was around 8.3 million metric tons in 2020-21, with an average yield of around 21 metric tons per hectare. Maharashtra is the largest producer of cabbage in India, accounting for over 20% of the country's total production.

Fig. 8.3: Cabbage crop

- Cabbage is an important crop in India, not only in terms of its nutritional value but also in terms of its economic importance. Cabbage is a rich source of vitamins, minerals, and antioxidants, making it a popular ingredient in many Indian dishes. It is also a major source of income for many farmers in India, particularly in the states of Maharashtra and Gujarat.
- The importance of cabbage in India can be gauged from the fact that it is one of the most widely traded vegetables in the country. The major markets for cabbage in India are Mumbai, Delhi, Kolkata, and Chennai, with the crop being transported from the major producing states to these markets.
- In addition to its nutritional and economic importance, cabbage also plays a significant role in Indian cuisine. It is used in a variety of dishes, including curries, soups, salads, and pickles. The crop is also used in traditional Indian medicine, with its leaves and roots being used to treat a variety of ailments.

Overall, cabbage is an important crop in India, with its production and consumption playing a significant role in the country's agricultural economy and food culture.

8.2.1. Agroclimatic Conditions Required for Cultivation

Cabbage is a cool-season crop that requires specific agroclimatic conditions to grow well. Here are the detailed requirements:

- **Temperature**: Cabbage requires a moderate temperature, ranging from 10°C to 25°C, for optimal growth. It can tolerate light frosts but is sensitive to extreme temperatures. Prolonged exposure to temperatures above 25°C can lead to premature bolting (flowering), while temperatures below 10°C can cause damage to the crop.

- **Moisture:** Cabbage requires consistent moisture, especially during the germination and head-forming stages. The ideal annual rainfall for cabbage cultivation is around 600-800 mm. However, cabbage can be grown using irrigation, and a well-distributed rainfall or supplemental irrigation is essential to prevent moisture stress.
- **Soil:** Cabbage prefers well-draining, fertile soils with a pH range of 6.0-7.0. The ideal soil texture is a mix of sand, silt, and clay, with a moderate to high organic matter content. Cabbage can tolerate a wide range of soil types but is sensitive to waterlogged or saline soils.
- **Day length**: Cabbage is a long-day crop, requiring a day length of 12-14 hours for optimal growth. However, some varieties can tolerate shorter day lengths.
- **Altitude**: Cabbage can be grown at a wide range of altitudes, from sea level to 2,000 meters above sea level. However, at higher elevations, the crop may require additional care, such as protection from frost and wind.
- **Wind protection**: Cabbage requires protection from strong winds, which can cause damage to the plants and disrupt the formation of tight heads. Windbreaks, such as rows of trees or shrubs, can help provide the necessary protection.
- **Sunlight**: Cabbage requires full sun to partial shade, depending on the variety. In warmer climates, some shade can help prevent bolting and reduce heat stress.
- **Growing season**: The ideal growing season for cabbage varies depending on the variety and climate. In general, cabbage is a cool-season crop that does best in the spring or fall, when temperatures are mild. In regions with mild winters, cabbage can be grown as a winter crop.

8.2.2. Optimum weather conditions required for sowing

For optimal growth and to prevent bolting (premature flowering) in cabbage, the ideal weather conditions for sowing are characterized by temperatures ranging from 18-22°C (64-72°F) for germination and 15-20°C (59-68°F) for growth.

- Consistent moisture, especially during the germination phase (7-10 days), is crucial, and well-draining soil is essential to prevent waterlogged conditions. Cabbage requires short days (10-12 hours of daylight) to prevent bolting, so it's best to sow seeds 8-10 weeks before the last expected frost date in your area.
- Additionally, fertile, well-draining soil with a pH between 6.0 and 7.0 is necessary for healthy growth. It's also important to provide wind

protection, as strong winds can cause physical damage and stress, leading to bolting.

By sowing cabbage seeds under these optimal conditions, you can minimize the risk of bolting and promote healthy growth.

8.2.3. Weather Conditions Favorable for High Productivity of Cabbage

For high productivity in cabbage cultivation, favorable weather conditions play a crucial role.

- Temperature is a critical factor, and optimal temperatures for cabbage growth range from 15°C to 20°C. This temperature range promotes healthy growth, encourages leaf expansion, and supports the development of dense, compact heads.
- During the germination phase, which typically lasts 7-10 days, consistent moisture is essential. The soil should be kept consistently moist, but not waterlogged, to prevent seed rot and ensure healthy seedling establishment.
- Adequate daylight is also necessary for high productivity in cabbage. Cabbage is a cool-season crop that thrives in short days with 10-12 hours of daylight. As the days lengthen, cabbage plants can become stressed, leading to bolting or premature flowering. Therefore, it's essential to plant cabbage at the right time, typically 8-10 weeks before the last expected frost date in your area.
- Soil quality also significantly impacts cabbage productivity. Cabbage prefers fertile, well-draining soil with a pH between 6.0 and 7.0. The soil should be rich in organic matter, which helps retain moisture, suppress diseases, and provide essential nutrients for healthy growth.
- Wind protection is another critical factor in cabbage cultivation. Strong winds can cause physical damage, stress the plants, and lead to bolting. Providing wind protection using row covers, windbreaks, or other forms of protection can help minimize these risks and promote healthy growth.

Overall, a combination of optimal temperatures, consistent moisture, adequate daylight, fertile soil, and wind protection can help create favorable weather conditions for high productivity in cabbage cultivation.

8.2.4. Effect of Extreme Weather Events

Extreme weather events can have a significant impact on cabbage cultivation, affecting crop yields, quality, and overall productivity. Here are some of the ways extreme weather events can affect cabbage cultivation:

- Prolonged periods of high temperatures, typically above 25°C, can cause cabbage plants to become stressed, leading to bolting or premature flowering. This can result in a loss of crop quality and reduced yields. High temperatures can also increase the risk of disease, such as downy mildew and powdery mildew, which can further compromise crop health.

- Drought conditions can severely impact cabbage cultivation, particularly during the germination and head formation stages. Insufficient moisture can cause seeds to germinate slowly or unevenly, leading to reduced stands and lower yields. Drought can also cause cabbage plants to become more susceptible to pests and diseases.
- Flooding and excessive rainfall can be detrimental to cabbage cultivation, causing waterlogged soil conditions that can lead to root rot and other diseases. Excess moisture can also promote the growth of weeds, which can compete with cabbage plants for water, nutrients, and light.
- Strong winds can cause physical damage to cabbage plants, breaking or bruising leaves and disrupting head formation. Wind can also increase the risk of disease by spreading pathogens and reducing the effectiveness of fungicides.
- Hailstorms can cause significant damage to cabbage crops, breaking or bruising leaves and disrupting head formation. Hail damage can also create entry points for diseases, further compromising crop health.
- Frost and freezing temperatures can cause significant damage to cabbage crops, particularly during the germination and head formation stages. Frost can damage or kill cabbage seedlings, while freezing temperatures can cause mature cabbage heads to become damaged or discolored.

To mitigate the effects of extreme weather events on cabbage cultivation, farmers can adopt several strategies, including:

- Using climate-resilient cabbage varieties that are bred to tolerate extreme weather conditions
- Implementing conservation agriculture practices, such as mulching and cover cropping, to reduce soil erosion and promote soil health
- Using irrigation systems to manage water stress during drought conditions
- Providing wind protection using row covers or windbreaks
- Monitoring weather forecasts to anticipate and prepare for extreme weather events
- Diversifying cabbage crops to spread risk and reduce reliance on a single crop.

By adopting these strategies, farmers can reduce the risks associated with extreme weather events and promote more resilient and sustainable cabbage cultivation practices.

8.3. Broccoli (*Brassica oleracea* var. italic)

Broccoli is a nutrient-rich vegetable that has gained popularity in India due to its numerous health benefits (Fig. 8.4). Here's an overview of broccoli's area, productivity, and importance in India:

Fig. 8.4: Broccoli crop

- India is not a major producer of broccoli globally, but it is grown in several states, primarily in the Himalayan region and in Maharashtra. As per the Indian Horticultural Database, the area under broccoli cultivation in India is approximately 12,000 hectares, with a production of around 150,000 metric tons.
- Broccoli is considered a "superfood" due to its high nutritional value, including vitamins C and K, fiber, and cancer-fighting compounds. Its importance in India can be attributed to several factors:
  - Health benefits: Broccoli is rich in antioxidants and has anti-inflammatory properties, making it beneficial for overall health and well-being.
  - Cancer prevention: The sulforaphane present in broccoli has been shown to have anti-cancer properties, making it a valuable addition to a healthy diet.
  - Economic benefits: Broccoli is a high-value crop, providing a lucrative income source for farmers, particularly in the Himalayan region.
  - Culinary uses: Broccoli is a versatile vegetable that can be used in various dishes, from soups and salads to stir-fries and curries.

Overall, broccoli is a valuable crop in India, offering numerous health benefits, economic advantages, and culinary uses.

8.3.1. Agroclimatic Conditions Suitable for Growing of Broccoli

Broccoli is a cool-season crop that requires specific agroclimatic conditions to grow. The ideal conditions for growing broccoli are characterized by cool temperatures, moderate humidity, and adequate moisture.

- **Temperature**: Broccoli requires a cool and moderate temperature regime. The optimal temperature for broccoli growth ranges from 15°C to 20°C. Temperatures above 25°C can lead to premature flowering, while temperatures below 10°C can cause bolting. A temperature range of 10°C to 15°C is ideal for seed germination and early growth.
- **Moisture**: Broccoli requires consistent and adequate moisture, especially during the germination and seedling stages. The crop requires about 600-800 mm of rainfall or irrigation per growing season. However, excessive moisture can lead to root rot and other diseases. Well-draining soils and proper irrigation management are essential to prevent waterlogged conditions.
- **Humidity** : Broccoli prefers a relatively high humidity, ranging from 60-80%. High humidity helps to prevent moisture stress and promotes healthy growth.
- **Day Length**: Broccoli is a long-day crop, requiring a day length of 12-14 hours to induce flowering. However, some varieties are bred to be day-neutral, allowing them to grow and produce flowers under a wide range of day lengths.
- **Soil:** Broccoli grows best in well-draining, fertile soils with a pH range of 6.0-7.0. The crop requires adequate nutrients, especially nitrogen, phosphorus, and potassium. Organic matter, such as compost or manure, can help improve soil fertility and structure.
- **Altitude**: Broccoli can be grown at a wide range of altitudes, from sea level to over 2,000 meters. However, the ideal altitude for broccoli cultivation is between 1,000-1,500 meters, where the temperature is cool and the humidity is moderate.
- **Growing Season:** The ideal growing season for broccoli is during the cooler months of the year. In India, the best time to plant broccoli is from September to November or from January to March, depending on the region and climate.

8.3.2. Optimum Weather Conditions Required for Sowing of Broccoli

The agroclimatic conditions required for sowing broccoli are quite specific. To ensure optimal germination and growth, the following conditions should be met:

- **Temperature:** The ideal temperature for sowing broccoli is between 10°C to 20°C. Temperatures above 25°C can lead to poor germination, while temperatures below 10°C can cause delayed germination. A temperature range of 15°C to 18°C is considered optimal for seed germination.
- **Soil Temperature**: Soil temperature is also an important factor to consider when sowing broccoli. The ideal soil temperature for broccoli germination is between 10°C to 15°C. Soil temperatures above 20°C can lead to poor germination, while temperatures below 5°C can cause seed rot.
- **Moisture:** Adequate moisture is essential for seed germination. The soil should be kept consistently moist during the germination period, which typically lasts around 7-10 days. However, excessive moisture can lead to seed rot and poor germination.
- **Soil Depth**: Broccoli seeds should be sown at a depth of about 1-2 cm. Sowing seeds too deeply can lead to poor germination, while sowing seeds too shallowly can cause them to dry out.
- **Sowing Time**: In India, the best time to sow broccoli depends on the region and climate. In general, broccoli can be sown in the winter months (September to November) or in the early spring months (January to March). Sowing broccoli during the hot summer months is not recommended, as high temperatures can lead to poor germination and growth.
- **Weather Conditions**: Weather conditions during the sowing period should be calm and settled. Avoid sowing broccoli during periods of heavy rainfall, strong winds, or extreme temperatures, as these conditions can lead to poor germination and growth.

8.3.3. Weather Conditions Favorable for High Productivity

To achieve high productivity in broccoli cultivation, the following agroclimatic conditions are favorable:

- **Temperature:** A moderate temperature range of 15°C to 20°C is ideal for broccoli growth and development. This temperature range promotes healthy plant growth, optimizes flower formation, and enhances curd quality.

- **Moisture:** Adequate and consistent moisture is essential for high broccoli productivity. The crop requires about 600-800 mm of rainfall or irrigation per growing season. Well-distributed rainfall or supplemental irrigation during critical growth stages can significantly improve yields.
- **Humidity :** A relative humidity of 60-80% is favorable for broccoli growth. High humidity helps to prevent moisture stress, promotes healthy growth, and reduces the incidence of pests and diseases.
- **Day Length:** Broccoli is a long-day crop, requiring a day length of 12-14 hours to induce flowering. However, some varieties are bred to be day-neutral, allowing them to grow and produce flowers under a wide range of day lengths.
- **Soil:** Well-draining, fertile soils with a pH range of 6.0-7.0 are ideal for broccoli cultivation. The crop requires adequate nutrients, especially nitrogen, phosphorus, and potassium. Organic matter, such as compost or manure, can help improve soil fertility and structure.
- **Altitude**: Broccoli can be grown at a wide range of altitudes, but the ideal altitude for high productivity is between 1,000-1,500 meters. At this altitude, the temperature is cool, and the humidity is moderate, promoting healthy growth and development.
- **Growing Season**: The ideal growing season for broccoli is during the cooler months of the year. In India, the best time to plant broccoli is from September to November or from January to March, depending on the region and climate. Extreme weather conditions can have a significant impact on broccoli cultivation, affecting its growth, development, and productivity.

8.3.4. Effect of Extreme Weather Conditions

- **High Temperatures**: High temperatures above 25°C can cause broccoli to bolt, leading to premature flowering and a reduction in curd quality. Prolonged exposure to high temperatures can also lead to sunscald, causing white or yellowish blotches on the curds. Additionally, high temperatures can increase the incidence of pests and diseases, such as aphids, whiteflies, and powdery mildew.
- **Low Temperatures**: Low temperatures below 10°C can cause broccoli to become dormant, leading to a delay in growth and development. Prolonged exposure to low temperatures can also cause broccoli to bolt, leading to premature flowering. Frost damage can also occur, causing damage to the leaves and curds.

- **Drought:** Drought can have a significant impact on broccoli productivity, causing a reduction in curd size and quality. Drought stress can also lead to an increase in the incidence of pests and diseases, such as aphids and powdery mildew. Broccoli requires consistent moisture, especially during the critical growth stages, to produce high-quality curds.
- **Excessive Rainfall:** Excessive rainfall can cause broccoli to become waterlogged, leading to root rot and other diseases. Excessive moisture can also lead to an increase in the incidence of pests, such as slugs and snails. Additionally, excessive rainfall can cause the curds to become discolored and develop an unpleasant texture.
- **Wind:** Strong winds can cause broccoli plants to become damaged, leading to a reduction in productivity. Wind can also cause the curds to become discolored and develop an unpleasant texture. Additionally, wind can increase the incidence of pests and diseases, such as aphids and powdery mildew.
- **Hail:** Hail can cause significant damage to broccoli plants, leading to a reduction in productivity. Hail can cause the leaves and curds to become damaged, leading to an increase in the incidence of pests and diseases. Additionally, hail can cause the curds to become discolored and develop an unpleasant texture.
- **Frost:** Frost can cause significant damage to broccoli plants, leading to a reduction in productivity. Frost can cause the leaves and curds to become damaged, leading to an increase in the incidence of pests and diseases. Additionally, frost can cause the curds to become discolored and develop an unpleasant texture.

The extreme weather conditions can have a significant impact on broccoli cultivation, affecting its growth, development, and productivity. It's essential for farmers to take measures to mitigate the effects of extreme weather conditions, such as using irrigation systems, row covers, and other forms of protection.

8.4. Rutabaga (*Brassica napus* - Napobrassica Gp)

Rutabaga, also known as knol-khol or navalkol, is a cross between a cabbage and a turnip (Fig. 8.5).

Fig. 8.5: Rutabaga crop

While it is not as widely cultivated or consumed in India as some other vegetables, it does have a presence in certain regions.

- In terms of area, rutabaga is primarily grown in the cooler climates of India, such as in the states of Jammu and Kashmir, Himachal Pradesh, and Uttarakhand. It is also grown to a lesser extent in other parts of the country.
- Regarding production, India is not among the top producers of rutabaga globally. However, it does produce a significant quantity to meet domestic demands. According to the Indian Council of Agricultural Research, the country produces around 100,000 metric tons of rutabaga annually.
- Rutabaga holds nutritional significance due to its richness in vitamins C and K, fiber, and antioxidants. It is also low in calories, making it a healthy addition to various dishes. In Indian cuisine, rutabaga is often used in soups, stews, and curries, and is also pickled or served as a vegetable dish.
- It is grown in Ladakh. Ladakh's cold climate and short growing season make it suitable for cultivating rutabaga and other hardy, cool-season crops.
- In Ladakh, rutabaga is typically planted in late spring to early summer, when the soil has thawed, and the risk of frost has decreased. The crop matures in about 3-4 months, and is usually harvested in late summer to early fall. Rutabaga is an important crop in Ladakh, where it is consumed locally and also sold in markets. It is valued for its nutrition.

8.4.1. Agroclimatic Conditions Required for Growing Rutabaga

Rutabaga is a cool-season crop that requires specific agroclimatic conditions to grow. It thrives in temperate climates with moderate temperatures, adequate moisture, and well-defined seasons.

- Temperature is a critical factor in rutabaga cultivation. The ideal temperature for germination is between 7°C and 10°C. As the crop grows, it prefers day time temperatures between 15°C and 18°C and nighttime temperatures around 10°C. Rutabaga is sensitive to extreme temperatures; prolonged exposure to temperatures above 25°C can lead to bolting, while temperatures below 0°C can cause damage or kill the crop.
- Moisture is another essential factor for rutabaga growth. The crop requires consistent and adequate moisture, especially during the germination and root development stages. Rutabaga needs about 600-800 mm of rainfall or irrigation per growing season. However, excessive moisture can lead to root rot and other diseases.
- Soil is also crucial for rutabaga cultivation. The crop prefers well-draining, loose soils with a pH between 6.0 and 7.0. Rutabaga can tolerate a wide range of soil types but grows best in fertile, sandy loams with good organic matter content.
- Rutabaga is typically grown at elevations ranging from sea level to 2,500 meters above sea level. At higher elevations, the crop benefits from the cooler temperatures and longer growing season.
- Day length and sunlight exposure are also important factors in rutabaga growth. The crop requires about 12-14 hours of daylight during the growing season. Adequate sunlight exposure promotes healthy growth, while insufficient sunlight can lead to weak and leggy plants.
- Wind protection is essential for rutabaga cultivation, especially in areas with strong winds. Rutabaga plants can be damaged or uprooted by strong gusts, reducing yields and affecting quality.

Overall, rutabaga requires a specific combination of temperature, moisture, soil, elevation, day length, and sunlight exposure to grow. By providing these conditions, farmers can successfully cultivate rutabaga and enjoy a bountiful harvest.

8.4.2. Weather Conditions Suitable for Sowing

The ideal weather conditions for sowing rutabaga are characterized by cool temperatures, adequate moisture, and minimal risk of extreme weather events.

- Temperature is a critical factor in determining the suitability of weather conditions for sowing rutabaga. The ideal temperature for sowing rutabaga is between 7°C and 10°C. At this temperature range, the seeds germinate quickly, and the seedlings establish themselves rapidly.

- Soil temperature is also important, and it should be around 10°C to 15°C for optimal germination. If the soil temperature is too high, above 20°C, it can lead to poor germination and reduced seedling vigor.
- Adequate moisture is essential for germination and seedling establishment. The soil should be consistently moist but not waterlogged. A light rain or irrigation before sowing can help to settle the soil and provide sufficient moisture for germination.
- Weather stability is also crucial for sowing rutabaga. Avoid sowing during periods of extreme weather fluctuations, such as during unseasonal heatwaves or cold snaps. Rutabaga is sensitive to temperature fluctuations, and extreme weather events can lead to poor germination, reduced seedling vigor, or even crop failure.
- In general, the best time to sow rutabaga is during the spring or early summer, when the weather is cool and stable. In regions with mild winters, rutabaga can also be sown in the fall, about 8 to 10 weeks before the first frost.
- It's also essential to avoid sowing rutabaga during periods of high wind, which can lead to soil erosion and reduced seedling establishment. Similarly, avoid sowing during periods of prolonged drought, as this can lead to poor germination and reduced seedling vigor.

Overall, the ideal weather conditions for sowing rutabaga are characterized by cool temperatures, adequate moisture, and minimal risk of extreme weather events. By sowing rutabaga during these conditions, farmers can optimize germination, seedling establishment, and ultimately, crop yields.

8.4.2. Weather Conditions Favorable for High Productivity

For optimal growth and high productivity of rutabaga, a specific set of weather conditions is favorable.

- Temperature is a critical factor, and rutabaga grows best in temperatures ranging from 15°C to 18°C during the day and around 10°C at night. This temperature range promotes healthy growth, root development, and sugar accumulation.
- Adequate moisture is also essential for high productivity. Rutabaga requires consistent and moderate soil moisture, especially during the root development stage. A rainfall of around 600-800 mm or supplemental irrigation during dry spells can help meet the crop's moisture needs.
- Rutabaga also benefits from a moderate level of sunshine. Full sun to partial shade is ideal, with at least 6 hours of direct sunlight per day. This helps promote photosynthesis, growth, and root development.

- Adequate soil temperature is also crucial for root growth and development. Soil temperatures between 10°C and 15°C are ideal for rutabaga growth.
- Wind protection is also important, as strong winds can damage the plants and reduce yields. A location with some natural wind protection, such as a hill or a group of trees, can help minimize wind damage.
- In terms of seasonal timing, rutabaga grows best in the cooler seasons of spring or fall. In regions with mild winters, it can also be grown during the winter months.

Overall, the favorable weather conditions for high productivity of rutabaga include moderate temperatures, adequate moisture, moderate sunshine, and wind protection. By providing these conditions, farmers can optimize growth, root development, and ultimately, yields.

8.4.3. Effect of Extreme Weather Conditions

Extreme weather conditions can have a significant impact on rutabaga crops, affecting their growth, development, and productivity.

- Prolonged periods of high temperatures, above 25°C, can lead to bolting, where the plant prematurely goes to seed, reducing its edible root quality. High temperatures can also cause sunscald, leading to discoloration and reduced market value.
- On the other hand, extreme cold temperatures, below 0°C, can cause damage or kill the crop. Rutabaga is sensitive to frost, and even light frosts can damage the leaves and roots.
- Drought or water stress can also significantly impact rutabaga crops. Insufficient moisture can lead to reduced growth, lower yields, and decreased root quality. Conversely, excessive rainfall or flooding can cause root rot and other diseases.
- Strong winds can damage the plants, reducing yields and affecting quality. Wind can also lead to soil erosion, reducing soil fertility and affecting future crops.
- Hailstorms can cause physical damage to the plants, leading to reduced yields and lower quality roots.
- Extreme weather conditions can also increase the susceptibility of rutabaga crops to pests and diseases. For example, warm and humid conditions can favor the development of fungal diseases, while dry and hot conditions can increase the incidence of insect pests.

Overall, extreme weather conditions can have devastating effects on rutabaga crops, highlighting the need for farmers to adopt climate-resilient practices and strategies to mitigate these impacts.

8.5. Conclusion

- Brassica group of vegetables, including broccoli, cauliflower, cabbage, kale, and rutabaga, are cool-season crops that require specific agroclimatic conditions to grow. Understanding the effects of temperature, moisture, sunlight, and wind on these crops is crucial for optimizing their growth, development, and productivity.
- By recognizing the favorable and unfavorable weather conditions for each crop, farmers and agricultural practitioners can adopt strategies to mitigate the negative impacts of extreme weather events and climate change.
- This knowledge can also inform breeding programs aimed at developing climate-resilient Brassica varieties. Ultimately, a thorough understanding of the agroclimatology of Brassica group of vegetables is essential for ensuring the sustainability and food security of these important crops.

9

Role of Weather on Mushroom Production

Mushrooms are a nutritious and versatile food commodity that has gained significant importance in India in recent years. With a growing demand for healthy and organic food options, mushroom production has emerged as a lucrative venture for Indian farmers (Fig. 9.1).

Fig. 9.1: Types of edible mushrooms

- India is home to a diverse range of wild and cultivated mushroom species, with over 200 varieties found across the country. The majority of mushroom production in India is concentrated in the states of Uttar Pradesh, Punjab, Haryana, and Himachal Pradesh.
- Mushroom production in India has witnessed significant growth over the past few decades, driven by advances in technology, improved farming practices, and increasing demand. According to the Indian Council of Agricultural Research, the country produces over 1.5 million metric tons of mushrooms annually, with a value of over ₹3,000 crores.

- Mushrooms are an important crop in India due to their high nutritional value, medicinal properties, and economic significance. They are an excellent source of protein, fiber, and various essential vitamins and minerals. Mushrooms are also rich in antioxidants and have been found to have potential health benefits, including reducing inflammation, improving cardiovascular health, and boosting the immune system.
- In addition to their nutritional and medicinal properties, mushrooms also play a significant role in India's economy. Mushroom cultivation provides employment opportunities for thousands of people, particularly in rural areas. The mushroom industry also generates significant revenue for the country through exports, with India exporting mushrooms to countries such as the United States, Europe, and the Middle East.
- The significance of mushrooms in India extends beyond their nutritional and economic value. Mushrooms also have cultural and traditional significance, with various species being used in traditional medicine and cooking. In some parts of India, mushrooms are considered a delicacy and are served at special occasions and festivals.
- In recent years, the Indian government has launched several initiatives to promote mushroom production and consumption in the country. These initiatives include providing financial support to mushroom farmers, establishing mushroom research and development centers, and promoting mushroom-based products and value chains.

Overall, mushrooms are an important crop in India, offering a range of nutritional, medicinal, and economic benefits. With the government's support and the growing demand for healthy and organic food options, the mushroom industry in India is poised for significant growth and development in the coming years. While it is technically possible to grow mushrooms in open fields, it is not a common or recommended practice for several reasons:

- Mushrooms are sensitive to temperature, humidity, and light, and open fields often have unpredictable and uncontrolled environmental conditions. This can lead to reduced yields, poor quality, and increased susceptibility to pests and diseases.
- Open fields are often exposed to direct sunlight, which can inhibit mushroom growth and cause damage to the fruiting bodies. Mushrooms typically require partial shade to full shade, depending on the species.
- Wind and rain can also damage mushroom crops in open fields. Strong winds can disrupt the delicate mushroom caps, while heavy rainfall can cause waterlogging and reduce mushroom quality.

- Weeds and other vegetation in open fields can compete with mushrooms for nutrients, water, and space, reducing yields and quality.
- Pests and diseases can also be more problematic in open fields, where mushrooms are more exposed to insects, animals, and pathogens.
- For these reasons, mushroom cultivation is often done in controlled environments, such as greenhouses, tunnels, or indoor growing facilities. These environments allow for more precise control over temperature, humidity, light, and other factors, resulting in higher yields, better quality, and reduced risks. However, there are some exceptions, such as:
  - Oyster mushrooms, which can be grown on logs or stumps in open fields.
  - Shiitake mushrooms, which can be grown on logs in shaded areas.
  - Some species of wild mushrooms, which can be found growing naturally in open fields.

In general, though, mushroom cultivation in open fields is not a common practice, and controlled environments are preferred for optimal growth and quality.

9.1. Facilities Required for Mushroom Production

Mushroom production requires a controlled environment with specific facilities to ensure optimal growth and quality. Here are some of the key facilities required for mushroom production:

- A climate-controlled growing room or house is essential for maintaining the optimal temperature, humidity, and light conditions for mushroom growth. The room should be well-insulated, draft-free, and equipped with heating, cooling, and ventilation systems.
- A sterilization room or area is necessary for sterilizing substrates, equipment, and other materials to prevent contamination. This room should be equipped with autoclaves, hot water baths, or other sterilization equipment.
- A substrate preparation room or area is required for preparing the growth substrate, which can include composting, pasteurizing, or sterilizing the substrate.
- A spawning room or area is necessary for inoculating the substrate with mushroom spawn. This room should be maintained at a high level of cleanliness and have a controlled environment.
- A growing bed or shelf system is required for supporting the mushroom bags or trays. The beds or shelves should be designed to allow for good air circulation, easy maintenance, and minimal contamination.

- Irrigation and misting systems are necessary for maintaining optimal humidity and moisture levels during the growing process.
- A cooling system is required for maintaining optimal temperatures during the growing process. This can include air conditioning units, evaporative cooling systems, or other cooling technologies.
- A pest and disease management system is essential for preventing and controlling pests and diseases that can affect mushroom growth and quality.
- A packaging and storage facility is necessary for preparing mushrooms for market and storing them at optimal temperatures and humidity levels.
- A laboratory or testing facility can be useful for monitoring and controlling the quality of the mushroom crop, as well as for conducting research and development activities.

Overall, the specific facilities required for mushroom production can vary depending on the scale and type of production, as well as the specific needs of the mushroom variety being grown.

9.2. Popular Varieties of Mushrooms

India is home to a diverse range of mushroom species, with several types being commercially cultivated and consumed. Here are some of the most popular types of mushrooms produced in India:

- **Button Mushrooms** (*Agaricus bisporus*): These are one of the most widely consumed varieties of mushrooms in India, known for their mild flavor and versatility in various dishes.
- **Oyster Mushrooms** (*Pleurotus ostreatus*): Also known as "dhingri" in Hindi, these mushrooms are popular for their delicate flavor and soft texture.
- **Milky Mushrooms** (*Calocybe indica*): These mushrooms are known for their bright white color and are widely cultivated in states like Karnataka, Tamil Nadu, and Kerala.
- **Shiitake Mushrooms** (*Lentinufa edodes*): Native to East Asia, these mushrooms are gaining popularity in India for their rich, smoky flavor and potential health benefits.
- **Reishi Mushrooms** (*Ganoderma lucidum*): Also known as "lingzhi" mushrooms, these are prized for their medicinal properties and are widely cultivated in India.
- **Straw Mushrooms** (*Volvariella volvacea*): These mushrooms are commonly known as "Chinese mushrooms" and are widely cultivated in states like Madhya Pradesh, Uttar Pradesh, and Maharashtra .

9.2.1. Button type Mushrooms

Button Mushroom is the most popular mushroom variety grown and consumed the world over (Fig. 9.2).

Fig. 9.2: Button type mushroom

In India, its production earlier was limited to the winter season, but with technology development, these are produced almost throughout the year in small, medium and large farms, adopting different levels of technology. The species being grown in most farms is the white button mushroom (*Agaricus bisporus*) belonging to Class Basidiomycetes and Family Agaricaceae.?

- Impact of weather factors
  - Temperature: A slight drop in temperature can trigger mushroom growth.
  - Humidity: High humidity helps to promote mushroom growth.
  - Light: A controlled light cycle can help regulate mushroom growth.
- **Environmental Control:** To optimize mushroom growth, cultivators carefully control the environment, including:
  - Temperature: Maintained between 55-60°F (13-15°C) for optimal mushroom growth.
  - Humidity: Maintained at 80-90% to promote healthy mushroom growth.
  - Light: Controlled to simulate natural daylight cycles, which helps regulate mushroom growth.
  - CO_2 levels: Maintained at optimal levels for mushroom growth, typically between 1000 to 2000 ppm.

9.2.2. Oyster Mushrooms

Oyster mushroom is popularly known as 'dhingri' in India and grows naturally in the temperate and tropical forests on dead and decaying wooden logs or sometimes on dying trunks of deciduous or coniferous woods. It may also grow on decaying organic matter (Fig. 9.3).

Fig. 9.3: Oyster mushroom

- The fruit bodies of this mushroom are distinctly shell or spatula shaped with different shades of white, cream, grey, yellow, pink or light brown depending upon the species.
- It is one of the most suitable fungal organisms for producing protein rich food from various agro-wastes or forest wastes without composting.

Agro-climatic Requirements

- Oyster mushroom can grow at moderate temperature ranging from 20 to 30 degree C and humidity 55-70% for a period of 6 to 8 months in a year.
- It can also be cultivated in summer months by providing the extra humidity required for its growth. In hilly areas above 900m. (m.s.l.), the best growing season is during March/April to September/October and in the lower regions from September/October to March/April.
- Successfully cultivate oyster mushrooms in a controlled environment, maintain a temperature range of 22-28°C (72-82°F) and a relative humidity of 80-90%.

Optimum Weather Conditions Required

- **Temperature:** Oyster mushrooms thrive in a moderate temperature range, with the optimal conditions being between 22-28°C (72-82°F). While they can tolerate slightly lower temperatures, maintaining this range is crucial for optimal growth and fruiting.

- **Humidity**: High humidity is essential for oyster mushroom cultivation, with a relative humidity of 80-90% being ideal. Adequate humidity helps prevent the substrate from drying out and ensures proper fruiting body development.
- **Light**: Oyster mushrooms need some light to grow, but they don't require direct sunlight. Provide indirect light or dim, artificial light to encourage fruiting.
- **Air Exchange**: Ensure good air exchange to prevent the buildup of carbon dioxide, which can negatively affect mushroom growth. Regular ventilation can help maintain optimal conditions for oyster mushroom cultivation.

9.2.3. Milky Mushrooms

Milky mushrooms are a type of edible fungi native to India. Milky mushrooms hold significant value in India, particularly for their culinary and nutritional benefits. They are prized for their unique flavor and texture, and are used in various Indian dishes, particularly in vegetarian cuisine (Fig. 9.4).

Fig. 9.4: Milky mushroom

These mushrooms are rich in protein, fiber, and various minerals, making them a nutritious addition to a balanced diet. Additionally, milky mushroom cultivation provides a livelihood for many rural communities in India.

- **Weather Controls Required for Quality and Productivity:** To cultivate high-quality milky mushrooms, specific weather conditions need to be maintained.
  - The ideal temperature for vegetative growth is between 20-25°C, while fruiting occurs optimally between 15-20°C. The temperature range for overall growth is between 10-30°C.
  - In terms of humidity, milky mushrooms require a humid environment, with an optimum humidity of 80-90% for vegetative growth and 90-

95% for fruiting. The humidity range for overall growth is between 60-95%.

- Low to moderate light intensity, ranging from 500-1,000 lux, is ideal for milky mushroom cultivation. The light duration should be around 10-12 hours per day.
- Maintaining consistent moisture levels and gentle air circulation are also crucial for healthy growth and optimal yields.

9.2.4. Shiitake Mushrooms

Shiitake mushrooms are a type of edible fungi native to East Asia. They are prized for their rich, smoky flavor and meaty texture (Fig. 9.5). Shiitake mushrooms are widely cultivated and are considered a delicacy in many cuisines.

Fig. 9.5: Shiitake mushroom

- Shiitake mushrooms hold significant value in various aspects. They are a popular ingredient in many Asian dishes, particularly in Chinese, Japanese, and Korean cuisine. They are also rich in protein, fiber, and various vitamins and minerals, making them a nutritious addition to a balanced diet. Additionally, Shiitake mushrooms have been found to have potential health benefits, including immune system support, anti-inflammatory properties, and antioxidant activity.
- Shiitake mushroom cultivation is also a significant industry in many countries, providing a source of income for many farmers and rural communities.
- Weather Controls Required for Quality and Productivity: To cultivate high-quality Shiitake mushrooms, specific weather conditions need to be maintained.
    - The optimum temperature for mycelium growth is between 10-20°C, while fruiting occurs optimally between 15-25°C. The temperature range for overall growth is between 5-30°C. Sudden temperature

changes should be avoided, as they can stress the mycelium and reduce yields.

- In terms of humidity, Shiitake mushrooms require a humid environment, with an optimum humidity of 70-90% for mycelium growth and 80-95% for fruiting. Consistent humidity levels should be maintained to prevent moisture stress and promote healthy growth.
- Low to moderate light intensity, ranging from 500-1,000 lux, is ideal for Shiitake mushroom cultivation. The light duration should be around 10-12 hours per day, with diffuse, indirect light provided to prevent overheating and promote healthy growth.
- Maintaining consistent moisture levels is also crucial, with a relative humidity of 70-90% optimal. Watering should be done every 2-3 days, depending on the humidity and temperature. Clean, filtered water should be used to prevent contamination and promote healthy growth.
- Finally, gentle air circulation should be maintained to prevent CO_2 buildup and promote healthy growth. Adequate air exchange is also necessary to maintain optimal CO_2 levels and prevent moisture buildup. By controlling these environmental factors, Shiitake mushroom cultivators can optimize the quality and productivity of their crop.

9.2.5. Reishi Mushrooms

Reishi mushrooms are a type of edible fungi native to Asia, Europe, and North America (Fig. 9.6). They are prized for their potential health benefits and are considered a delicacy in many cuisines.

Fig. 9.6: Reishi mushroom

- Reishi mushrooms hold significant value in various aspects. They have been used in traditional medicine for centuries, particularly in Chinese and Japanese cultures, to promote health and well-being.
- Reishi mushrooms are rich in bioactive compounds, including polysaccharides, peptides, and triterpenoids, which have been found to have potential health benefits, including immune system support, anti-inflammatory properties, and antioxidant activity.
- Reishi mushroom cultivation is also a growing industry, providing a source of income for many farmers and rural communities.
- **Weather Controls Required for Quality and Productivity**: To cultivate high-quality Reishi mushrooms, specific weather conditions need to be maintained. The optimum temperature for mycelium growth is between 20-25°C, while fruiting occurs optimally between 15-20°C. The temperature range for overall growth is between 10-30°C. Sudden temperature changes should be avoided, as they can stress the mycelium and reduce yields.
    - In terms of humidity, Reishi mushrooms require a humid environment, with an optimum humidity of 70-90% for mycelium growth and 80-95% for fruiting. Consistent humidity levels should be maintained to prevent moisture stress and promote healthy growth.
    - Low to moderate light intensity, ranging from 500-1,000 lux, is ideal for Reishi mushroom cultivation. The light duration should be around 10-12 hours per day, with diffuse, indirect light provided to prevent overheating and promote healthy growth.
    - Maintaining consistent moisture levels is also crucial, with a relative humidity of 70-90% optimal. Watering should be done every 2-3 days, depending on the humidity and temperature. Clean, filtered water should be used to prevent contamination and promote healthy growth.
    - Finally, gentle air circulation should be maintained to prevent CO_2 buildup and promote healthy growth. Adequate air exchange is also necessary to maintain optimal CO_2 levels and prevent moisture buildup. By controlling these environmental factors, Reishi mushroom cultivators can optimize the quality and productivity of their crop.

9.2.6.: Straw Mushrooms

Straw mushrooms are a type of edible fungi native to Asia. They are widely cultivated and are considered a delicacy in many cuisines (Fig. 9.7).

Fig. 9.7: Straw mushroom

- Straw mushrooms hold significant value in various aspects. They are a good source of protein, fiber, and various vitamins and minerals, making them a nutritious addition to a balanced diet. Straw mushrooms are also rich in antioxidants and have been found to have potential health benefits, including immune system support and anti-inflammatory properties.
- Additionally, straw mushroom cultivation is a significant industry in many countries, providing a source of income for many farmers and rural communities.
- **Weather Controls Required for Quality and Productivity:** To cultivate high-quality straw mushrooms, specific weather conditions need to be maintained.
- The optimum temperature for mycelium growth is between 25-30°C, while fruiting occurs optimally between 20-25°C. The temperature range for overall growth is between 15-35°C. Sudden temperature changes should be avoided, as they can stress the mycelium and reduce yields.
- In terms of humidity, straw mushrooms require a humid environment, with an optimum humidity of 80-90% for mycelium growth and 90-95% for fruiting. Consistent humidity levels should be maintained to prevent moisture stress and promote healthy growth.
- Low to moderate light intensity, ranging from 500-1,000 lux, is ideal for straw mushroom cultivation. The light duration should be around 10-12 hours per day, with diffuse, indirect light provided to prevent overheating and promote healthy growth.
- Maintaining consistent moisture levels is also crucial, with a relative humidity of 80-90% optimal. Watering should be done every 2-3 days, depending on the humidity and temperature. Clean, filtered water should be used to prevent contamination and promote healthy growth.

Finally, gentle air circulation should be maintained to prevent CO2 buildup and promote healthy growth. Adequate air exchange is also necessary to maintain optimal CO2 levels and prevent moisture buildup. By controlling these environmental factors, straw mushroom cultivators can optimize the quality and productivity of their crop.

- **Additional Weather Considerations:** Straw mushroom cultivation is typically done outdoors, and weather conditions can significantly impact yields. Weather factors such as rainfall, wind, and extreme temperatures can damage crops and reduce yields. To mitigate these risks, straw mushroom cultivators often use greenhouses or other forms of protection to control the environment and ensure optimal growing conditions.

9.3. Conclusion

- Mushroom cultivation is a complex process that requires careful attention to detail and a controlled environment. The significance of mushroom cultivation extends beyond its culinary and nutritional value, as it also provides a source of income for many farmers and rural communities. However, mushroom cultivation is highly dependent on weather conditions, and fluctuations in temperature, humidity, light, and air circulation can significantly impact yields and quality.
- Therefore, implementing effective weather controls is crucial for successful mushroom cultivation. By maintaining optimal temperature, humidity, and light levels, and ensuring adequate air circulation, mushroom cultivators can optimize the quality and productivity of their crop. Additionally, using greenhouses, climate-controlled environments, or other forms of protection can help mitigate the risks associated with outdoor cultivation and ensure a consistent and high-quality yield.
- In conclusion, mushroom cultivation is a delicate process that requires careful attention to weather conditions. By understanding the specific weather requirements for different types of mushrooms and implementing effective weather controls, cultivators can optimize their yields, improve quality, and contribute to a sustainable and profitable mushroom industry.

10

Weather Effects on Other Vegetables

In the preceding chapters, we embarked on a comprehensive journey through the realm of vegetables, covering a wide range of species that belong to different botanical and culinary groups. These chapters provided an in-depth exploration of the characteristics, growth habits, and environmental requirements of various vegetables, as well as the effects of weather conditions on their productivity.

However, the diversity of vegetables is vast, and there are several other important species that were not covered in the earlier chapters. These vegetables may be regionally specific, possess unique characteristics, or have distinct environmental requirements. Despite their importance, they have remained outside the scope of our previous discussions.

This chapter aims to fill this knowledge gap by providing an overview of these additional vegetables, with a focus on understanding the impact of weather conditions on their growth, development, and productivity. By examining the effects of temperature, moisture, light, and other environmental factors on these vegetables, we can gain valuable insights into their cultivation requirements and optimize their production.

Through this chapter, we hope to provide a more comprehensive understanding of the vegetable kingdom, highlighting the diversity of species, their environmental requirements, and the strategies for maximizing their yields. By expanding our knowledge of these lesser-covered vegetables, we can promote their cultivation, conservation, and sustainable use, ultimately contributing to global food security and nutrition.

10.1. Moringa (*Moringa oleifera*)

Moringa is a tropical plant that thrives in a wide range of agroclimatic conditions, making it an ideal crop for cultivation in various regions (Fig. 10.1). To grow moringa successfully, the following agroclimatic conditions are required:

10.1.1. Agroclimatic conditions required

Temperature is a crucial factor in moringa cultivation.

- The ideal temperature range for moringa growth is between 25°C and 30°C. However, the plant can tolerate temperatures as high as 40°C and as low as 10°C for short periods. Prolonged exposure to temperatures above 40°C or below 10°C can lead to reduced growth and productivity.
- Moringa requires adequate moisture, especially during the germination and seedling stages. The ideal annual rainfall for moringa cultivation is between 500 mm and 1500 mm. However, the plant can tolerate drought conditions and can survive with minimal rainfall. In areas with low rainfall, irrigation is necessary to ensure optimal growth and productivity.

Fig. 10.1: Moringa as vegetable

- Moringa can thrive in a wide range of soil types, from sandy to clayey soils. However, the plant prefers well-drained soil with a pH range of 6.0 to 7.0. Soils with high salinity or alkalinity can lead to reduced growth and productivity. In areas with poor soil quality, organic amendments and fertilizers can be used to improve soil fertility and structure.
- Moringa requires full sun to partial shade, with a minimum of 6 hours of direct sunlight per day. In areas with intense sunlight, providing some shade can help prevent scorching of the leaves.
- Altitude is not a significant constraint for moringa cultivation. The plant can thrive at elevations ranging from sea level to 1000 meters above sea level. However, at higher elevations, the plant's growth rate and productivity may be reduced.
- Wind protection is necessary for moringa cultivation, especially in areas with strong winds. Moringa trees can be prone to wind damage, which can lead to reduced growth and productivity. Providing wind protection using windbreaks or shelterbelts can help mitigate this risk.

Overall, moringa is a highly adaptable plant that can thrive in a wide range of agroclimatic conditions. With proper care and management, moringa can be cultivated successfully in various regions, providing a valuable source of nutrition, income, and employment.

10.2. Baby Corn (*Zea mays*)

Baby corn, also known as young corn or sweet corn, is a popular and nutritious vegetable crop in India. The country is one of the largest producers of baby corn in the world, with the crop being grown in almost every state (Fig. 10.2).

- In terms of area under cultivation, baby corn is grown in over 200,000 hectares of land across India. The major baby corn-producing states in the country include Maharashtra, Madhya Pradesh, Uttar Pradesh, and Bihar. These states account for over 60% of the country's total baby corn production.

Fig. 10.2: Baby corn as vegetable

- India is the second-largest producer of baby corn in the world, accounting for over 10% of global production. The country produces over 2.5 million tons of baby corn every year, with the majority being consumed domestically. The production of baby corn is a significant contributor to the country's economy, providing employment and income opportunities for thousands of farmers and rural communities.
- Baby corn holds great significance in Indian cuisine and culture. The vegetable is a popular ingredient in many traditional dishes, including curries, soups, and salads. Baby corn is also used in various forms, including fresh, frozen, and canned, making it a versatile and convenient ingredient for households and restaurants.
- The importance of baby corn in India extends beyond its culinary significance. The crop is an excellent source of nutrition, providing a rich source of vitamins, minerals, and antioxidants. Baby corn is rich in vitamins C and E, as well as minerals like potassium and folate. The

vegetable is also low in calories and high in fiber, making it an ideal ingredient for health-conscious consumers.

- In recent years, there has been a growing demand for baby corn in India, driven by increasing awareness of the vegetable's health benefits and culinary versatility. This has led to the establishment of a number of baby corns processing units and manufacturing facilities across the country, providing employment opportunities for thousands of people.
- The Indian government has also recognized the potential of baby corn as a valuable crop, and has initiated several programs and schemes to promote its cultivation and production. These initiatives include providing subsidies and incentives to farmers, establishing baby corn research and development centers, and promoting the export of baby corn products.

Overall, baby corn is a highly significant and important crop in India, providing a range of benefits and opportunities for farmers, rural communities, and the country as a whole.

10.2.1. Agroclimatic Conditions Required

Baby corn, being a warm-season crop, requires specific agroclimatic conditions to complete its life cycle. The life cycle of baby corn can be divided into several stages, including germination, seedling establishment, vegetative growth, flowering, and harvesting.

During the germination stage, baby corn requires a warm and moist environment. The ideal temperature for germination is between 20°C and 30°C, with an optimal temperature of around 25°C. The soil should be kept consistently moist during this stage, with a relative humidity of 60-80%. The germination period typically lasts around 7-10 days.

- After germination, the seedlings require adequate light, temperature, and moisture to establish themselves. The ideal temperature for seedling establishment is between 22°C and 28°C, with an optimal temperature of around 25°C. The soil should be kept consistently moist, but not waterlogged, during this stage. The relative humidity should be around 60-80%. The seedling establishment period typically lasts around 1-2 weeks.
- During the vegetative growth stage, baby corn requires a warm and sunny environment. The ideal temperature for vegetative growth is between 25°C and 35°C, with an optimal temperature of around 30°C. The crop requires adequate moisture during this stage, with a relative humidity of 60-80%. The vegetative growth period typically lasts around 3-4 weeks.

- The flowering stage of baby corn is critical, as it determines the yield and quality of the crop. The ideal temperature for flowering is between 25°C and 30°C, with an optimal temperature of around 28°C. The relative humidity should be around 60-80% during this stage. The flowering period typically lasts around 1-2 weeks.
- Finally, the harvesting stage of baby corn requires careful planning to ensure optimal quality and yield. The ideal temperature for harvesting is between 20°C and 25°C, with an optimal temperature of around 22°C. The relative humidity should be around 60-80% during this stage. Baby corn is typically harvested around 60-70 days after sowing, when the ears are still soft and the kernels are still milky.
- In terms of rainfall, baby corn requires adequate moisture throughout its life cycle. The ideal annual rainfall for baby corn is between 600 mm and 1200 mm, with an optimal rainfall of around 900 mm. However, the crop can be grown successfully under irrigation, provided that the soil is well-drained and has a good water-holding capacity.

Overall, baby corn requires a warm and sunny environment, with adequate moisture and a well-drained soil, to complete its life cycle. The ideal agroclimatic conditions for baby corn are typically found in the tropical and subtropical regions of India, where the crop is widely grown.

10.3. Stem Vegetables

Stems used as vegetables, also known as stem vegetables, are an integral part of Indian cuisine and culture. These vegetables are obtained from the stems of various plants, which are harvested at different stages of maturity (Fig. 10.3). Stem vegetables are a significant component of Indian cooking, and their importance extends beyond their culinary value.

Fig. 10.3: Stem vegetables

- In India, stem vegetables have been consumed for centuries, with various regions having their own unique preferences and preparations. These vegetables are not only a good source of essential nutrients but also add texture, flavor, and variety to Indian dishes. Some of the most commonly consumed stem vegetables in India include bamboo shoots, lotus stems, and sugarcane and banana pith including asparagus and celery also.
- The significance of stem vegetables in India can be attributed to their nutritional value, versatility, and cultural importance. These vegetables are rich in vitamins, minerals, and antioxidants, making them a healthy addition to Indian meals. Stem vegetables are also incredibly versatile, with various cooking methods and preparations used to bring out their unique flavors and textures.
- In Indian cuisine, stem vegetables are used in a variety of dishes, from soups and stews to curries and salads. Bamboo shoots, for example, are commonly used in northeastern Indian cuisine, where they are stir-fried with spices and herbs. Lotus stems, on the other hand, are often pickled or used in salads, adding a crunchy texture and sweet flavor to dishes.
- The cultural importance of stem vegetables in India is also noteworthy. In many Indian communities, these vegetables are considered a delicacy and are often served at special occasions and festivals. For example, in some parts of India, sugarcane is considered a sacred plant and is offered to deities during festivals.
- In addition to their cultural and culinary significance, stem vegetables also play an important role in Indian agriculture and economy. Many farmers in India cultivate stem vegetables as a cash crop, providing them with a source of income and livelihood. The export of stem vegetables, such as bamboo shoots and sugarcane, also contributes to India's agricultural export earnings.

Overall, stem vegetables are a vital component of Indian cuisine, culture, and economy. Their nutritional value, versatility, and cultural significance make them an integral part of Indian cooking and traditions. As India continues to evolve and grow, the importance of stem vegetables is likely to remain unchanged, with these vegetables continuing to play a vital role in Indian cuisine and culture.

10.4. To Recap

- Agroclimatology plays a vital role in understanding the complex relationships between climate, weather, and vegetable crop production.

As the global population continues to grow, ensuring food security and sustainability has become a pressing concern. Vegetable crops, being an essential component of human nutrition, require careful consideration of agroclimatic factors to optimize their production. This book has highlighted the importance of agroclimatology in understanding the responses of vegetable crops to various climatic and weather factors, such as temperature, precipitation, solar radiation, and wind. By examining the impact of these factors on vegetable crop growth, development, and yield, farmers, researchers, and policymakers can make informed decisions to improve crop productivity, reduce losses, and promote sustainable agriculture practices.

- However, there is still a significant need for further research, particularly under Indian conditions, where the diverse climate and agro-ecological regions pose unique challenges. India's vast geographical area encompasses a range of climates, from tropical to temperate, and from humid to arid. This diversity of climates, combined with the country's rapidly growing population and increasing demand for food, makes it essential to develop location-specific agroclimatic models and decision-support systems. Climate change, rising temperatures, and changing precipitation patterns are expected to impact vegetable crop production in India, making it crucial to investigate the potential impacts of these changes on crop yields, water requirements, and pest and disease dynamics.
- By advancing our understanding of agroclimatology and its application in vegetable crop production, we can improve crop yields, enhance food security, and promote sustainable agriculture practices in India and beyond. This knowledge can be used to develop climate-resilient crop varieties, optimize irrigation schedules, and design effective pest and disease management strategies. Furthermore, agroclimatic information can be used to develop early warning systems for weather-related disasters, such as droughts, floods, and heatwaves, which can help farmers for taking proactive measures to mitigate losses. Ultimately, the application of agroclimatology in vegetable crop production has the potential to contribute significantly to India's food security, economic growth, and sustainable development goals.

Bibliography

I. Books/Reports/Chapters

- Fageria, M.S., Dhaka, R.S. and Choudhary, B.R. 2020. Vegetable Crops Production Technology. Vol - II. Kalyani Publishers. ISBN: 9789327268713. 252 pages
- Kumar S, Srivastava S & Verma S. 2024. Climate Change and Vegetable Production: Challenges and Solutions. String Production, Uttrakhand. 326 pages
- Ramana Rao B.V., Surender Singh, V. Uma Maheswara Rao. 2025. Preparatory Microclimatology. Publisher: Professional Prints. New York USA. ISBN: 9781966695189 Pp 167.
- Ramana Rao B.V., Surender Singh, V. Uma Maheswara Rao. 2025. Conceptual Meteorology. Publisher: New India Publishing Agency, New Delhi. ISBN 978-9358874501 Pp 152.
- Ramana Rao B.V., Surender Singh, V. Uma Maheswara Rao. 2025. Primer on Agrohydrometeorology. Publisher: Professional Prints. New York USA. ISBN: 978-1966695301 Pp 144.
- Ramana Rao B.V., Surender Singh, V. Uma Maheswara Rao, Anu Bulusu. 2025. Farming Weather: Harnessing Citizen Science for Smart Agriculture. Publisher: Professional Prints. New York USA. ISBN: 978-1966695561. Pp 154.
- Ramana Rao B.V., Surender Singh, V. Uma Maheswara Rao. 2025. Forecasting the Future: Effective Climate Guidance on Monsoons for Farm Prosperity. Publisher: Professional Prints. New York USA. ISBN: 978-1966695912. Pp 246.
- Ramana Rao B.V., Surender Singh, V. Uma Maheswara Rao. 2025. Preparatory Microclimatology. Publisher: Professional Prints. New York USA. ISBN: 978-1966695189 Pp 167.
- Ramana Rao B.V., Surender Singh, V. Uma Maheswara Rao. 2025. Conceptual Meteorology. Publisher: New India Publishing Agency, New Delhi. ISBN 978-9358874501 Pp 152.
- Ramana Rao B.V., Surender Singh, V. Uma Maheswara Rao. 2025. Primer on Agrohydrometeorology. Publisher: Professional Prints. New York USA. ISBN: 978-1966695301 Pp 144.
- Ramana Rao B.V., Surender Singh, V. Uma Maheswara Rao, Anu Bulusu. 2025. Farming Weather: Harnessing Citizen Science for Smart Agriculture. Publisher: Professional Prints. New York USA. ISBN: 978-1966695561. Pp 154.
- Ramana Rao B.V., Surender Singh, V. Uma Maheswara Rao. 2025. Forecasting the Future: Effective Climate Guidance on Monsoons for Farm Prosperity. Publisher: Professional Prints. New York USA. ISBN: 978-1966695912. Pp 246.
- Solankey, S.S. and Kumari, M. 2023. Advances in Research on Vegetable Production Under a Changing Climate Vol. 2. Springer Publication. ISBN: 9783031208393. 369 pages

- Stoddart. K.2024. The Climate Change–Resilient Vegetable Garden: How to Grow Food in a Changing Climate. Cool Springs Press Publisher. ISBN. 9780760384732. 208 pages.
- Welbaum, G.E.2015. Vegetable production and practices. CABI Publisher. ISBN : 978-1-845938024476. 451 pages

II. Journals/Research Papers/Proceedings

- Agricultural and Forest Meteorology. https://www.sciencedirect.com/journal/agricultural-and-forest-meteorology
- Chinese Journal of Agrometeorology. https://zgnyqx.ieda.org.cn/EN/1000-6362/home.shtml
- Duhan, A. Vegetable Crops: Risks and Losses Faced by Farmers. *Advances in Research*. 12(6): 1-8.
- Dumitru, E.A.; Berevoianu, R.L.; Tudor, V.C.; Teodorescu, F.-R.; Stoica, D.; Giucă, A.; Ilie, D.; Sterie, C.M. 2023. *Climate Change Impacts on Vegetable Crops*: A Systematic Review. Agriculture. 13,1891.
- Italian Journal of Agrometeorology. https://riviste.fupress.net/index.php/IJAm/issue/view/15
- Journal of Agril Meteorology.:https://journals.indexcopernicus.com/search/details?id=46231
- Journal of Agrometeorology.https://journal.agrimetassociation.org/index.php/jam/issue/view/73
- MAUSAM. https://mausamjournal.imd.gov.in/index.php/MAUSAM
- Rajasekar, M.; Arumugam, T. and Ramesh Kumar, S.2013. Influence of weather and growing environment on vegetable growth and yield. *Journal of Horticulture and Forestry*. 5(10), pp. 160-167.
- Shaban D. Abou-Hussein. 2012. Climate Change and Its Impact on the Productivity and Quality of Vegetable Crops. *Journal of Applied Sciences Research*. 8(8): 4359-4383.
- Spaldon, S., Samnotra, R.K. and Chopra, S. 2015. Climate resilient technologies to meet the challenges in vegetable production. *International Journal of Current Research and Academic Review*. 3(2): 28-47.

III. Web Resources

- https://www.agrifarming.in/growing-vegetables-in-summer-a-full-guide
- https://www.indiamart.com/proddetail/vegetables-crops-24663328830.html
- https://link.springer.com/chapter/10.1007/978-3-031-37424-1_4
- https://www.fao.org/newsroom/detail/latest-ipcc-report-highlights-the-critical-need-to-transform-agrifood-systems-as-a-way-to-mitigate-and-adapt-to-climate-change/en
- https://www.niubol.com/Product-knowledge/What-isAgrometeorology.html
- https://www.youtube.com/watch?app=desktop&v=0vz4vy7Z4lg& pp=ygUJI3NlZHN0b3Jp
- https://www.hortibiz.com/newsitem/news/study/importance-of-climate-and-context-alongside-yield-prediction-in-tomato-production-
- https://www.apnikheti.com/en/pn/agriculture/horticulture/vegetable-crops/chilli
- https://www.geturbanleaf.com/blogs/vegetables/how-to-grow-your-own-pepper-outdoors
- https://www.gardenia.net/plant/persea-americana-avocado-tree
- https://hyofarmsindia.com/collections/creeper-vegetables
- https://urbangreenguide.com/english-cucumber-plant/

- https://eruvaaka.com/our-agriculture/ridge-gourd-cultivation/
- https://www.ekrishikendra.com/home/sen/t-93
- https://kisanvedika.bighaat.com/growing-organic-bottle-gourd-a-planting-guide/
- https://blog.nurserylive.com/2016/09/05/9617-and-gardening-in-india
- https://www.agrifarming.in/coccinia-farming-income-kovakkai-ivy-gourd#goog_rewarded
- https://krishijagran.com/agripedia/pumpkin-cultivation-and-practices/
- http://www.celkau.in/crops/vegetables/Ashgourd/ashgourd.aspx
- https://www.trustbasket.com/blogs/how-to-grow/tinda-vegetable-round-gourd-facts-benefits-cultivation-how-to-grow-at-home
- https://www.asiafarming.com/parwal-farming-pointed-gourd-techniques
- https://www.snapdeal.com/product/sponge-gourd-satputiya-seeds-for/619620545951
- https://goldenhillsfarm.in/products/beans-rajma-kidney-bean-creeper-seeds/
- https://www.stocksandgreen.com/products/climbing-french-bean-blue-lake-seeds
- https://www.agrifarming.in/cluster-bean-cultivation-income-guar-project-report
- https://dailytrust.com/experts-outline-guidelines-as-cowpea-farmers-set-for-cultivation/
- https://www.gardenersworld.com/how-to/grow-plants/how-to-grow-broad-beans/
- https://www.freshpoint.com/news/root-vegetables-vs-tubers/
- https://dehaat.in/en/blog/Potato-How-to-make-the-tuber-bigger
- https://wikifarmer.com/library/en/article/sweet-potato-soil-requirements-soil-preparation-and-planting
- https://www.agrifarming.in/elephant-foot-yam-cultivation
- https://agritech.tnau.ac.in/horticulture/horti_vegetables_tapioca.html
- https://www.britannica.com/plant/taro-plant
- https://wikifarmer.com/library/en/article/carrot-soil-requirements-soil-preparation-and-planting
- https://extension.sdstate.edu/beets-how-grow-it
- https://www.apnikheti.com/en/pn/agriculture/horticulture/vegetable-crops/radish
- https://www.freepik.com/premium-photo/turnip-greens-plants-cultivation-field-detail-leaves-turnip-greens-cultivation_85256954.htm
- https://www.asiafarming.com/green-peas-cultivation
- https://www.farmatma.in/onion-cultivation/
- https://agricultureguruji.com/garlic-farming/
- https://www.farmatma.in/ginger-cultivation-grow-ginger/
- https://www.crossfit.com/essentials/leafy-greens-the-ultimate-brain-food
- https://www.apnikheti.com/en/pn/agriculture/horticulture/vegetable-crops/spinach
- https://www.tropicalpermaculture.com/amaranth-plant.html
- https://www.wildcraftia.com/plant/fenugreek/
- https://plantura.garden/uk/vegetables/malabar-spinach/malabar-spinach-overview
- https://www.gardenersworld.com/how-to/grow-plants/sorrel-grow-guide/
- https://www.apnikheti.com/en/pn/agriculture/horticulture/medicinal-plants/pudina
- https://www.agrifarming.in/kale-cultivation
- https://www.agrifarming.in/earn-excellent-income-returns-from-lettuce-farming
- https://gardenerspath.com/plants/vegetables/grow-collard-greens/
- https://gardenerspath.com/plants/vegetables/grow-watercress/
- https://www.gardeningknowhow.com/edible/herbs/curry/growing-curry-leaf.htm
- https://www.apnikheti.com/en/pn/agriculture/horticulture/spice-and-condiments/coriander
- https://www.nutritionadvance.com/allium-vegetables/

- https://www.garden.eco/growing-shallots
- https://www.gardenersworld.com/how-to/grow-plants/how-to-grow-leeks/
- https://www.sustainablemarketfarming.com/2015/05/11/garlic-scapes-three-weeks-to-bulb-harvest/
- https://gardenerspath.com/plants/vegetables/grow-scallions/
- https://www.allthatgrows.in/blogs/posts/grow-brassica-plant
- https://www.rhs.org.uk/vegetables/cauliflower/grow-your-own
- https://www.yara.co.uk/crop-nutrition/cabbage/
- https://eos.com/blog/how-to-grow-broccoli/
- https://gardenerspath.com/plants/vegetables/grow-rutabaga/
- https://www.onmanorama.com/food/features/2021/06/06/best-edible-mushrooms-in-india-across-the-world.html
- https://www.pashudhanpraharee.com/step-by-step-procedure-for-mushroom-farming-in-india/
- https://www.fieldforest.net/product/reishi-table-top-farm-instruction-sheet/instruction-sheets
- https://www.picturethisai.com/care/temperature/Moringa_oleifera.html
- https://eng.kisanofindia.com/agriculture-success-stories/baby-corn-farming-padmashree-farmer-kanwal-singh-chouhan-formula
- https://housing.com/news/stem-vegetables/

Subject Index